Two Manuals Combined into One Book

Medical Platoon Leader's Handbook

And

Combat and Operational Stress Control

Department of Defense

Pentagon
Publishing

FIELD MANUAL
NUMBER 4-02.4

HEADQUARTERS
DEPARTMENT OF THE ARMY
Washington, DC, 24 August 2001

MEDICAL PLATOON LEADERS' HANDBOOK
TACTICS, TECHNIQUES, AND PROCEDURES

TABLE OF CONTENTS

DISTRIBUTION RESTRICTION: Approved for public release; distribution is unlimited.

---

*This publication supersedes FM 8-10-4, 16 November 1990.

PREFACE

This field manual (FM) provides information on the structure and operation of all medical platoons and medical sections that are organic to combat and combat support (CS) battalions and squadrons. It is directed toward the medical platoon leader and medical platoon members. The tactics, techniques, and procedures (TTP) provided are not all-inclusive. They provide a way of performing a particular mission, but may require modification based on mission, enemy, terrain, troops, time available, and civilian considerations (METT-TC).

This publication provides information on the organization of the division and how medical platoons and sections organic to division units provide combat health support (CHS). It outlines the responsibilities of medical platoon/section leaders. It provides definitive information on planning, rehearsing, and conducting CHS at Echelon I. It provides TTP for directing, controlling, and managing CHS at the medical platoon/section level. It describes the troop-leading procedures for CHS operations and identifies interface and coordination requirements with other brigade medical elements. This publication provides doctrine for the Force XXI medical platoon.

This FM is not a stand-alone reference. It is a doctrine publication that speaks to the current Army of Excellence (AOE) Division and the Force XXI Digitized Division and Brigade CHS and will require the user to be familiar with FMs 4-02.21, 8-10, 8-10-1, 8-10-3, 8-10-5, 8-10-6, 8-10-7, 8-10-9, 8-10-26, 8-42, and 8-55.

This publication implements the following North Atlantic Treaty Organization (NATO) Standardization Agreements (STANAGs) and American, British, Canadian, and Australian (ABCA) Quadripartite Standardization Agreements (QSTAGs):

Title	STANAG	QSTAG
Medical and Dental Supply Procedures	2128	
Regulations and Procedures for Road Movement and Identification of Movement Control and Traffic Control Personnel and Agencies	2454	
Orders for the Camouflage of the Red Cross and the Red Crescent on Land in Tactical Operations	2931	
Medical Materiel Management during Patient Evacuation		435
Minimum Labelling Requirements for Medical Materiel		436

As the Army Medical Department (AMEDD) transitions to the 91W military occupational specialty (MOS), positions for 91B and 91C will be replaced by 91W when new unit modification table(s) of organization and equipment (MTOE) take effect.

Users of this publication are encouraged to submit comments and recommendations to improve the publication. Comments should include the page, paragraph, and line(s) of the text where the change is recommended. The proponent for this publication is the United States (US) Army Medical Department Center and School (AMEDDC&S). Comments and recommendations should be forwarded directly to **Commander, AMEDDC&S, ATTN: MCCS-FCD-L, 1400 East Grayson Street, Fort Sam Houston, Texas 78234-5052**, or by using the e-mail addresses on the Doctrine Literature website at *http://dcdd.amedd.army.mil/index1.htm* (click on Doctrine Literature).

Unless this publication states otherwise, masculine nouns and pronouns do not refer exclusively to men.

Use of trade or brand names in this publication is for illustrative purposes only and does not imply endorsement by the Department of Defense (DOD).

CHAPTER 1

COMBAT ORGANIZATIONS

Section I. THE DIVISION

1-1. Background

The division is the largest Army organization that trains and fights as a tactical team. Largely self-sustaining, it is capable of independent operations. The division is a unit of maneuver organizations with varying numbers and types of combat, CS, and combat service support (CSS) units. It may be armored, mechanized, light infantry, airborne, or air assault; each can conduct operations over a wide range of environments. The success of Army operations depends on the success of its divisions.

1-2. Role of the Division

Traditionally, divisions have operated as part of a US corps. In corps operations, divisions are normally comprised of 9 to 12 maneuver battalions, organic artillery battalions, and supporting CS and CSS units. Divisions perform a wide range of tactical missions and, for limited periods, are self-sustaining. The corps augments divisions, as missions require. All divisions must be able to deploy and conduct offensive and defensive operations, stability operations, and support operations.

Section II. TYPES OF DIVISIONS

1-3. Armored and Mechanized Divisions

a. Army of Excellence. The US Army's armored and mechanized divisions (referred to collectively as heavy divisions) provide mobile, armor-protected firepower. Armored and mechanized divisions are normally employed for their mobility, survivability, lethality, and psychological effects on the enemy. These divisions destroy enemy armored forces. They can seize and secure land areas and key terrain. During offensive operations, armored and mechanized divisions can rapidly concentrate over-whelming lethal combat power to break or envelop enemy defenses or offensive formations. These divisions then continue the attack to destroy fire support, command and control (C2), and logistics elements. Their mobility allows them to rapidly concentrate, attack, reinforce, or block enemy forces. Their collective protection systems enable them to operate in a nuclear, biological, and chemical (NBC) environment. Armored and mechanized divisions operate best in open terrain where they gain the advantage with their mobility and long-range, direct-fire weapons. Because of strategic lift requirements, armored and mechanized forces are slow to deploy from home or staging bases into an area of operations (AO). They have high consumption rates of supplies, can deploy relatively few dismounted infantry, and have limited use in restrictive terrain. See FM 71-100 for organizational structure of the division.

b. Force XXI, Digitized Division. The Army's Force XXI Division is a redesign of our current divisions that represent a leap forward into the realm of 21st Century technology. The smaller Force XXI Division possesses greater lethality and quicker mobility, as well as the CSS imperative of situational understanding. The digital technological enablers will enhance situational understanding and provide the means for information dominance by enabling friendly forces to share a complete common relevant picture (CRP). This provides the commanders a CRP while communicating and targeting in real or near real-time. Digitization permits the division to conduct operations over an extended battle space by increasing the operational areas of responsibility for all maneuver elements. Digitization will decrease decision-making time by optimizing the flow of information. This information enables Force XXI commanders to quickly mass forces allowing the division to defeat a larger, but less technologically advanced enemy. It will contribute increased lethality, survivability, and operational tempo while reducing the potential for fratricide.

1-4. Light Infantry Division

The light infantry division fights as part of a larger force in conventional conflicts and conducts missions as part of a joint force in stability operations and support operations. The light infantry division is one of our most rapidly and strategically deployable divisions. Its C2 structure readily accepts any augmentation forces, permitting task organizing for any situation. The factors of METT-TC largely determine the augmentations required for the division. The optimum use of light forces is as a division under corps control, its mission capitalizing on its capabilities. The division exploits the advantages of restricted terrain and limited visibility. It achieves mass through the combined effects of synchronized small-unit operations and fires, rather than through the physical concentration of forces on the battlefield. Light division forces physically mass only when risk to the force is low and the payoff is high. The division deploys as an entity; widely dispersed to conduct synchronized, but decentralized, operations primarily at night or during periods of limited visibility. Light force limitations include their austere CS and CSS systems, and their requirement for support from the corps or joint force headquarters, based on METT-TC. For organizational structure of the light infantry division, see FM 71-100.

1-5. Airborne Division

The airborne division can rapidly deploy anywhere in the world to seize and secure vital objectives. The airborne division must be able to conduct forced entry operations. It conducts parachute assaults to capture initial lodgments, execute large-scale tactical raids, secure intermediate staging bases or forward operating bases for ground and air operations, or rescue US nationals besieged overseas. It also can serve as a strategic or theater reserve as well as reinforcement for forward presence forces. The airborne division can assault deep into the enemy's rear areas to secure terrain or interdict enemy supply and withdrawal routes. It can seize and repair airfields to provide a forward operating base and airheads for follow-on air-landed forces. It is capable of all other missions assigned to light infantry divisions. The airborne division uses its strategic and operational mobility to achieve surprise on the battlefield. The US Air Force (USAF) can accurately deliver the airborne division into virtually any objective area under almost any weather condition. All equipment is air transportable; most are air-droppable. All personnel are trained for parachute assaults and airborne operations. Engagements with enemy armored or motorized formations require special consideration. The division does not have sufficient armored protection to defeat heavier armored formations at close range. Antitank weapons in the division compensate for, but do not completely offset, this deficit. For division organizational structure, see FM 71-100.

1-6. Air Assault Division

The air assault division combines strategic deployability with tactical mobility within its AO. It attacks the enemy deep, fast, and often over extended distances and terrain obstacles. The air assault division must be able to conduct forced entry operations. The airmobile division of the Vietnam era provided the US Army the operational foundation, experience, and tactics for today's air assault operations. Air assault operations have evolved into combat, CS, and CSS elements (aircraft and troops) deliberately task-organized for tactical operations. Helicopters are completely integrated into ground force operations. Air assault operations generally involve insertions and extractions under hostile conditions, as opposed to mere air movement of troops to and from secure locations about the battlefield. Once deployed on the ground, air assault infantry battalions fight like battalions in other infantry divisions; however, normal task organization of organic aviation results in greater combat power and permits rapid aerial redeployment. The rapid tempo of operations over extended ranges enables the division commander to rapidly seize and maintain the tactical initiative. For division organizational structure, see FM 71-100.

1-7. Medium Division

The medium division consists of one armored brigade, one mechanized brigade, and one air assault brigade, and traditional heavy division aviation, CS, and CSS units. The Army designed this division to provide commanders operational flexibility with armor lethality and light infantry strength in restrictive terrain.

Section III. THE DIVISION STAFF

1-8. The Division Commander

The division commander is responsible for everything the division does. He assigns missions, delegates authority, and provides guidance, resources, and support to accomplish the mission.

1-9. Assistant Division Commanders

Within a division there are two assistant division commanders (ADC). The division commander prescribes their duties, responsibilities, and relationships with the staff and subordinate units. Normally, the responsibilities are broken down as operations and training (or maneuver) and support. Light divisions have an ADC for operations and training (ADC-OT), while in the heavy division, he is referred to as the ADC for maneuver (ADC-M) and, in both heavy and light divisions, there is an ADC for support (ADC-S).

1-10. Chief of Staff

The chief of staff directs the efforts of both the coordinating and special staffs. His authority usually amounts to command of the staff.

1-11. Staff Section

The command sergeant major (CSM), Assistant Chief of Staff, G1 (Personnel), Assistant Chief of Staff, G2 (Intelligence), Assistant Chief of Staff, G3 (Operations and Plans); and Assistant Chief of Staff, G4 (Logistics) function at division level in much the same way their counterpart staffs function at battalion and brigade level. The Assistant Chief of Staff, G5 (Civil Affairs) is the civil-military operations officer. This position is normally authorized only at division level and higher. For a detailed discussion of staff organization and functions, see FM 101-5. Provided in paragraph 1-33 is a detailed discussion of the coordinating staff and, in paragraph 1-34, a detailed discussion of the special staff.

Section IV. ORGANIZATIONS OF DIVISIONS

1-12. Division Organizations

All divisions are generally organized with a similar basic design. This design comprises a division headquarters and headquarters company (HHC), three ground maneuver brigades, an aviation brigade, an artillery brigade (referred to as division artillery [DIVARTY]), a support command, a cavalry squadron, an air defense artillery (ADA) battalion, an engineer battalion or brigade, a signal battalion, a military intelligence battalion, a military police (MP) company, and in most cases, a chemical company. The division headquarters provides C2 for the division's organic, attached, or supporting units. The HHC provides logistics support and personnel for the division headquarters and staff sections. Ministry teams in each division unit provide religious support to soldiers and their families. These teams provide worship opportunities, pastoral care, religious education, and spiritual fitness. The HHC is normally located near the division's main command post (CP). See FM 71-100 for definitive information on division organizations.

1-13. Ground Maneuver Brigade

The maneuver brigade headquarters provides the C2 facilities necessary to employ maneuver and fires. The only unit permanently assigned to the brigade is the brigade HHC. The necessary combat, CS, and CSS units to accomplish the brigade mission are attached, under operational control (OPCON), or placed in support of the brigade. The brigade's HHC furnishes logistics support (including equipment and personnel) and security for the brigade headquarters staff sections. The brigade normally controls from two to five attached maneuver battalions. It can be employed in autonomous or semiautonomous operations when properly organized for combat. Field Manuals 7-30 and 71-3 contain details on employment of the brigade. Ground maneuver battalions and additional units are placed in a command relationship to the brigade headquarters. This allows the division to accomplish missions in any environment. As units are added to brigades and the division, the division support command (DISCOM) is modified to meet changes in the division's supply, maintenance, and medical requirements.

1-14. Aviation Brigade

The aviation brigade is a maneuver force of organic, attached, and supporting Army aviation units. They include attack, air assault, reconnaissance (RECON), electronic warfare (EW), and general support

(GS) units. The division and aviation brigade commanders can tailor the brigade for virtually any combat, CS, and CSS operation to accomplish division missions. The brigade is most effective when its aerial forces concentrate at critical times or places to destroy units and exploit enemy vulnerabilities. The brigade extends the division capability to simultaneously strike the enemy throughout his depth and from multiple directions. The aviation brigade commander may be required to operate over great distances with his forces spread throughout the division's AO. This makes timely and accurate coordination difficult. Coordination is one the most important functions of the aviation brigade staff. Corps aviation brigades may augment or support the division. Aviation units in these organizations include attack helicopter and assault helicopter battalions, medium helicopter and theater aviation companies, and command aviation battalions. Field Manual 1-111 is the doctrinal base for aviation brigade operations.

1-15. Division Artillery

The DIVARTY is the division's primary organic indirect fire support organization. It normally comprises cannon and rocket artillery. Fire support systems neutralize, suppress, or destroy enemy forces. The DIVARTY provides close support, interdiction, and counterfire fire support to division operations. The division normally receives additional field artillery (FA) support from the corps or joint force commander. This support could include cannon and multiple launch rocket system (MLRS) battalions or brigades. Corps 155-millimeter cannons and MLRS battalions are organized similar to DIVARTY battalions. Corps artillery battalions have a liaison section to coordinate with supported units. Field Manual 6-20-2 provides further detail concerning DIVARTY operations.

1-16. Engineer Brigade

Divisional engineers accomplish mobility, countermobility, survivability, general engineering, and geospatial missions. Additionally, they may perform infantry combat missions and tasks when required. Armored and mechanized divisions have an organic engineer brigade; light, airborne, and air assault divisions have only an engineer battalion organic to the division. A division, when fully committed, normally requires a corps combat engineer battalion and a CS equipment company to augment its organic engineer units. Corps provides additional engineer units based on an METT-TC analysis. The airborne, light, and air assault divisions have limited hauling and earth-moving capabilities. This reduces their capacity for obstacle creation and reduction, protective shelter construction, and combat route missions (mobility, counter-mobility, and survivability). The engineer brigade commander (the engineer battalion commander in light divisions) serves as the division engineer. He coordinates the efforts of all engineers working within the division sector. He requests support from higher headquarters based on his engineer estimate and the commander's concept for the operation. Additional assets may be attached, under OPCON, or in direct support (DS) of the division. Corps engineer units often reinforce a division. Field Manual 5-71-100 details division engineer operations.

1-17. Division Support Command

The DISCOM provides division-level CSS to all organic assigned and attached elements of the division. It furnishes limited CSS to nondivisional units in the division area. The DISCOM routinely performs the functions of arming, fueling, fixing, moving, and sustaining soldiers and their systems. The DISCOM commander's role is complex. He is a brigade-level commander and the division's principal CSS operator. He exercises full command authority over organic units in the support command. He also has a close relationship with the division G4 and the ADC-S because of their overlapping interests. Although the division G4 has coordinating staff responsibility for logistics planning and develops division-level plans, policies, and priorities, the DISCOM commander advises the division staff during the formulation of plans, estimates, policies, and priorities. The ADC-S, on the other hand, commands and supervises all rear area operations. The G3, with the G4, and the DISCOM commander normally locate the CSS elements in the division rear area. The forward support battalions (FSB) provide DS to brigades and are positioned in the brigade support areas (BSA). Assigned to the FSB is the forward support medical company (FSMC) that provides Echelon II medical care to those battalions with organic medical platoons. The company provides both Echelon I and Echelon II medical treatment on an area support basis to units operating in the BSA and brigade AO. The FSMC establishes treatment facilities (division clearing stations) in the BSA. The FSMC performs the following functions:

- Treatment of patients with disease and nonbattle injuries (DNBI), triage of mass casualties, advanced trauma management (ATM), initial resuscitation and stabilization, and evacuation of patients incapable of returning to duty from battalion aid stations (BAS) to the FSMC.

- Ground evacuation for patients from BAS and designated collecting points to the FSMC.

- Sick call services for the BSA and brigade rear area.

- Maintenance of field health records for personnel receiving their primary care from the FSMC according to Army Regulation (AR) 40-66.

- Operational dental care (includes emergency and essential dental care).

- Emergency Class VIII resupply to unit operating in the brigade AO.

- Unit-level medical maintenance.

- Diagnostic medical laboratory and radiology services commensurate with Echelon II medical treatment facilities (MTF).

- Outpatient consultation services for patients referred from Echelon I MTF.

- Limited reinforcement and augmentation to supported medical platoons.

- Patient holding for up to 40 patients (20 patients in light infantry divisions) able to return to duty (RTD) within 72 hours.

- Reconstitution/regeneration support for maneuver medical platoons

The main support battalion (MSB) is located in the division support area (DSA) to provide area support to divisional units in the DSA and backup support to the FSB. Assigned to the MSB is the main support medical company (MSMC). The MSMC provides Echelons I and II medical care to units without organic medical treatment elements operating in the DSA. The MSMC establishes its MTF and base of operations in the DSA and—

- Provides advice and guidance to the MSB commander and his staff on the health of the command and CHS activities.

- Performs triage, initial resuscitation and stabilization, and preparation for evacuation of sick, injured, and wounded personnel.

- Provides medical evacuation (ground) support on an area support basis in the DSA. (Corps ambulances are normally used to evacuate patients from the BSA; however, this mission could be assigned to the MSMC.)

- Provides treatment squads that may operate independently of the division clearing station for limited time periods.

- Provides reconstitution support for forward deployed medical companies/elements.

- Provides sick call services for the BSA and brigade rear area.

- Maintains the field health records of personnel receiving their primary care from the FSMC according to AR 40-66.

- Provides operational dental care (includes emergency and essential dental care).

- Provides Class VIII resupply through the division medical supply office (DMSO) to division and nondivisional units via supply point distribution, logistical packages, redirected ambulance backhaul, and emergency deliveries.

- Performs unit-level medical maintenance.

- Provides diagnostic medical laboratory and radiology services commensurate with Echelon II MTF.

- Provides outpatient consultation services for patients referred from Echelon I MTF.

- Provides patient holding for up to 40 patients (20 patients in light infantry divisions) able to RTD within 72 hours.

- Provides mental health (MH) and neuropsychiatric consultation services and combat stress control (CSC) for the division.

- Provides preventive medicine (PVNTMED) and environmental health surveillance, inspections and consultation services for division units.

- Provides optometry support limited to eye examinations, spectacle assembly using presurfaced single-vision lenses, and repair services.

Medical elements from the BSA and DSA may be deployed forward with a forward logistics element (FLE) and temporarily provide support for forward areas. Additionally, corps support organizations may use FLE for special support requirements and to rapidly resupply as far forward as possible. Although the division has its own organic CSS units, it relies on corps and corps support command (COSCOM) units to sustain the division for continuous operations. (It may also coordinate through the civil affairs [CA] staff for available host-nation [HN] support.) The medical brigade supporting the corps provides CHS that includes hospitals in GS of the division. In DS, it provides a forward surgical team (FST), air and ground medical evacuation (MEDEVAC), CSC, and veterinary, PVNTMED, and combat health logistics (CHL) services that include Class VIII resupply and blood management. See FMs 63-2, 63-2-2, 63-20, 63-20-1, 63-21, and 63-21-1 for information on DISCOM CHS operations. See FMs 4-02.1, 4-02.21, 8-10-1, 8-10-3, 8-10-5, 8-10-9, 8-10-25, 8-10-26, and 8-51 for definitive information on CHS for the division.

1-18. Division Cavalry Squadron

The division cavalry squadron performs RECON and security for division operations. This helps the division commander to maneuver his brigades and battalions and attack the enemy at the most critical points. It fills gaps between units and establishes physical contact with divisional units and adjacent units. The division cavalry squadron also—

- Facilitates the division's movement with RECON, establishing contact points and passage points, and coordinating with higher and adjacent headquarters.

- Performs RECON and security operations in the division's rear area.

- Performs damage control and combat operations in the division's rear area when tasked as, or as part of, a tactical combat force (TCF).

See FM 17-95 for definitive information on cavalry squadron operations.

1-19. Air Defense Battalion

The division ADA battalion retains the division's freedom of maneuver, protects critical division assets, destroys enemy aircraft before they release their ordnance, and denies the enemy aerial RECON. The division's ADA battalion commander is the air defense coordinator. He integrates the division's air defense plan with both corps and echelons above corps (EAC) air defenses and integrates any OPCON air defense units. See FM 44-100 for definitive information on air defense operations and procedures.

1-20. Military Intelligence Battalion

The military intelligence (MI) battalion provides the division dedicated intelligence, counterintelligence (CI), and EW support, including communications intercept, direction finding capability, and electronic attack.

The MI battalion responds to the G2- and G3-assigned intelligence, CI, and EW requirements and missions. The MI battalion receives, analyzes, and disseminates intelligence information from echelons above division (EAD) as well as from assets within the division. The division MI battalion conducts G2-tasked intelligence collection operations according to the division commander's priority intelligence requirements and information requirements through the analysis and control element (ACE). The ACE is organic to the MI battalion and operates under the direction of the G2. Additionally, the MI battalion conducts G3-tasked EW operations missions per the commander's EW priorities. The MI battalion coordinates and directs corps and EAC intelligence and electronic warfare (IEW) assets supporting the division. A MI brigade performs IEW support at the corps level. It provides GS to the corps and, on occasion, may reinforce the division's intelligence efforts. When required, intelligence, imagery, EW, operations security (OPSEC) support, and interrogation assets directly support the divisions. See the FM 34-series (MI) manuals that provide the foundations for MI operations.

1-21. Signal Battalion

Signal support to the division is a collective and integrated application of communications, automation, and information services and systems. The signal battalion uses three distribution systems to cover the division's AO. They are the area common user system, the combat net radio (CNR) system, and the automatic data distribution system. Their integrated operation forms the division communications system. Signal units attached to or supporting the division will be under OPCON of the division signal battalion commander unless otherwise assigned or attached. The corps signal brigade links the division communications system to corps and higher echelons. The resulting architecture provides deployed forces with secure global, voice, and digital data communications. If required for special missions, corps and EAC signal assets can augment the division's existing signal capabilities. Specific responsibilities for the signal staff and the signal battalion are contained in FMs 11-50 and 101-5.

1-22. Military Police Company

The division MP company performs five functions in support of division operations:

- Maneuver and mobility support, to include support to river crossing operations, breaching operations, and passage of lines: straggler and dislocated civilian control; route RECON and surveillance; and main supply route (MSR) enforcement.

- Area security, to include RECON operations, area damage control, base/air base defense, response force operations and critical site, assets, or high-risk personnel security.

- Internment and resettlement operations, to include enemy prisoner of war (EPW) and civilian internee operations, confinement of US military prisoners and populace, and resource control.

- Law-and-order operations, to include law enforcement, criminal investigation, and US military prisoner confinement.

- Police intelligence operations, to include criminal information used to support the intelligence preparation of the battlefield (IPB), police assessment process, and joint information.

The MP company commander normally collocates with the division rear CP in the DSA to direct the use of his MP unit. Finally, the division normally receives one corps MP company to support sustained operations. Like the divisional MP company, this attached corps MP company works under the staff supervision of the provost marshal. Field Manual 3-19.1 details MP operations.

1-23. Division Chemical Company

The division chemical company is found in all divisions except light infantry. It reduces the effects of enemy NBC weapons and counters enemy sensor systems by using smoke and obscurants on division combat operations. Its primary focus is NBC RECON, decontamination (less patient decontamination), and smoke generation. Several NBC organizations and detachments provide added NBC defensive capabilities to the division. Corps chemical units may support corps CS and CSS units operating in the division area and may reinforce the divisional chemical company. Other sources of chemical support for divisions are mechanized or motorized smoke units, NBC RECON units, NBC decontamination units, and dual-purpose (light divisions only) smoke and decontamination companies. When additional companies are attached to or under OPCON of a division, a separate chemical battalion headquarters and headquarters detachment may be needed for command, control, and communications (C3). It will report directly to the division chemical officer for operational taskings. Field Manual 3-100 provides the doctrinal base for chemical company operations.

1-24. Division Band

The mission of the Army bands is to provide music to enhance unit cohesion and morale and to musically support military operations. Bands provide important support to information operations and should be integrated into public affairs, CA, and psychological operations (PSYOP) plans. Army bands are assigned secondary missions (except the US Army Band, the US Army Field Band, and the US Army Military Academy Band which have unique missions). These principal secondary missions are augmentation of security operations for CPs, or augmentation of perimeter security for EPW/civilian internee operations.

Section V. MANEUVER BRIGADES AND REGIMENTS

1-25. Organization of the Armored or Mechanized Infantry Brigade

The armored or mechanized infantry brigade is a combination of armored and mechanized infantry battalion task forces (TF) and other supporting units grouped under the command of a brigade headquarters. It participates in division or corps operations according to the principles and concepts set forth in FMs 71-100 and 100-5.

a. Divisional Brigades.

(1) Close combat-heavy brigades are the major subordinate maneuver commands of armored and mechanized infantry divisions. The only permanent unit assigned to a brigade is its HHC. The HHC provides direction and control over units assigned to, attached to, or supporting the brigade. The new Force XXI digitized brigade has gone through some organizational changes; for example, the maneuver battalion now has only three companies. There was also a redesign of its CSS in the maneuver battalions and the supporting DISCOM FSB. The Force XXI maneuver battalion with its enhanced computers, communications, and digitalization has real-time situational understanding that provides personnel a CRP.

(2) Divisional infantry, armored, and mechanized battalions are attached to brigades to destroy the enemy and to seize and hold terrain. Normally, each brigade can control three or four maneuver battalions with their CS and CSS units. When it is necessary to concentrate forces, control of more battalions may be necessary. However, the battalions assigned to a brigade must be limited to a number that can be controlled in a very complex battle situation.

(3) Light infantry battalions may be attached to the heavy brigade for specific missions and for a short duration. Use of light forces requires careful consideration of key employment and logistics support.

(4) The divisional brigade does not have support assets assigned to it; however, a habitual support relationship does exist between a designated FSB and the maneuver brigade. Normally, brigade support is also provided by a DS FA battalion; an ADA battery; an engineer company; a forward area signal platoon; an MP platoon; combat IEW elements; a tactical air control party (TACP); and a DISCOM FSB. Attack helicopter units may also operate with the brigade. When sorties are allocated for planning, USAF tactical air operations support the brigade.

b. Separate Mechanized Infantry and Armored Brigades.

(1) Since separate brigades conduct operations under corps command, they are organized to provide their own support. Units organic to the separate brigade include—

- A brigade HHC to provide C2 and limited CS assets, to include MP, chemical, and air defense elements.

- Tank and mechanized battalions to fight battles, destroy or disrupt enemy forces, and seize and hold terrain.

- An armored cavalry troop for RECON, security, and economy-of-force operations.

- A DS FA battalion to provide fire support.

- An engineer company for combat engineer support.

- An MI company to assist in collecting, processing, and disseminating intelligence and to support EW operations.

- A support battalion organized to provide CSS in the same way as the DISCOM's FSB provides CSS to divisional brigades, but with the added ability to link directly with the COSCOM for augmentation.

(2) Additional combat, CS, and CSS units may be attached to a separate brigade as required by the brigade's mission and operating circumstances. The separate brigade may be attached to a division (less support) but is usually controlled by a corps.

1-26. Organization of Infantry Brigades

a. Divisional Brigades.

(1) Light infantry, airborne, or air assault brigades are the major subordinate maneuver commands of infantry, airborne, or air assault divisions.

(2) There may be three or four brigades assigned to an infantry division, depending on operational requirements but in most cases it is three.

(3) Combat support and CSS are provided to the *brigade* by the *division.* Normally, FA support is provided by a light FA DS battalion. An engineer company, a forward area signal center platoon, combat IEW elements, and DISCOM forward support elements also routinely support a brigade. From time to time, attack helicopter units and USAF bombers may operate in support of the brigade.

b. Separate Light Infantry Brigades.

(1) Since separate brigades sometimes conduct independent operations, they are organized to provide their own support. Each is generally organized with—

- A brigade HHC to provide C2.

- Infantry battalions to destroy the enemy and to seize and hold terrain.

- A support battalion with several support units to provide CSS.

- A combat IEW company to assist in collecting, processing, and disseminating intelligence and to support EW operations.

- A light FA battalion to provide fire support.

- An engineer company for combat engineer support.

- An armored cavalry troop for RECON, security, and economy of force operations.

(2) Additional combat, CS, and CSS units may be attached to the separate brigade, as required. The separate brigade may be attached to a division or placed under the control of a higher command such as a corps.

1-27. Armored Cavalry Regiments

The armored cavalry regiments (ACR) are designed as either an ACR (armored) or an ACR (light) (L).

a. Armored Cavalry Regiment (Armored). The ACR is a self-contained combined arms organization composed of armored cavalry squadrons (ACS), an aviation squadron, a support squadron, and separate CS companies and batteries. The ACR is a separate unit that supports the corps or a joint task force (JTF). It is often reinforced by corps CS units and divisional maneuver battalions. The ACR operates independently over a wide area and at extended distances from other units. The ACR is a highly mobile, armored force capable of fighting the fully mechanized threat in the environmental states of war or conflict. The ACR may be rapidly deployed to a theater of operations by sealift. When supporting a light corps, limitations may exist in corps support capabilities, strategic mobility, and terrain restrictions. The regimental ACS, a highly mobile, armor-protected force, is discussed below. The regimental aviation squadron (RAS) provides the regiment with combat aviation assets. It is organized with air cavalry/RECON troops, attack helicopter troops, and an assault helicopter troop. The squadron adds a very responsive, terrain-independent combat capability to the regiment. The maneuverability and flexibility of the RAS enhance the combat flexibility of the regiment. The RAS may operate independently of or in close coordination with the ACS, or it may provide troops to the ACS.

b. Armored Cavalry Regiment (L). The ACR (L) is a self-contained combined arms organization capable of being packaged and rapidly deployed by air or sealift as part of a force projection Army responding rapidly to worldwide contingencies. The role of the ACR (L) may be traditional, initial entry, or follow-on. The traditional role would support a US corps or TF through a RECON, security, and economy-of-force capability. As an initial entry force, the ACR (L) would support Army or JTF operations with credible force as a demonstration of US resolve. In the follow-on role, the ACR (L) will follow an opposed entry force (division-ready brigade type) to expand the point of entry, to provide RECON and security, and to serve as the initial combat-capable maneuver force. Because of the C2 structure and support base within the regiment, it is a very modular organization capable of rapidly integrating armored forces into its task organization. This factor supports the Army with a force package that can be tailored for the situation and expanded once it is deployed to a theater if the situation dictates the need for armored forces. The ACR (L) is composed of ACS, an aviation squadron, a support squadron, and separate CS companies and batteries. The ACR (L) is a separate unit that supports the contingency corps. It is often reinforced by corps CS units and divisional maneuver battalions. The ACR (L) operates independently over a wide area and at extended distances from other units. The ACR (L) is a highly mobile force capable of executing the full range of doctrinal cavalry missions against a comparable threat in the environmental states of war or conflict. It is also capable of conducting stability operations and support operations. When supporting a light corps, limitations may exist in corps support capabilities, strategic mobility, and terrain restrictions. The organizational structure for the ACR (L) is similar to the ACR (armored) with some significant exceptions. These differences may require modification of the TTP prescribed for the ACR according to FM 17-95. The following assets or capabilities are not organic to the ACR (L):

- Neither the ACR (armored) nor the ACR (L) is authorized a tactical CP.
- The chemical company of the ACR (L) is not organized with a smoke platoon.
- The engineer company of the ACR (L) is not organized with bridging assets.
- Organic assets do not include digital terrain database development.
- The MI company of the ACR (L) does not have ground surveillance radar.
- The aviation squadron of the ACR (L) does not have attack helicopter troops.

Limited ballistic protection offered by the high-mobility multipurpose wheeled vehicle (HMMWV) and lack of organic tank assets require judicious application of standard cavalry doctrine. The ACR (L) is ideally suited for force packaging and employment by the contingency corps against a comparably equipped threat, but must be reinforced to defeat a modern mechanized or armored force. For definitive information on both the ACR (armored) and the ACR (L), see FM 17-95.

Section VI. THE BATTALION

1-28. Organization of the Infantry Battalion

a. Organization. An infantry battalion is organized and equipped to give it the capabilities needed to accomplish its missions. It is large enough to engage enemy regiments using a full range of organic and nonorganic weapons and support. However, it is small enough that the battalion commander can personally lead and immediately influence the action of his units in battle.

(1) To understand the organizational structure of the battalion, one must understand the organizational roles of echelons above and below the battalion and how the battalion serves as the interface for these echelons.

(2) Within the context of organizational roles, platoons normally fight as part of a company. Companies fight using their subordinate platoons as fire or maneuver elements. Battalions provide support to the companies; ensure the battlefield has depth; and synchronize the various arms and services to achieve the maximum effect from the available forces. The brigade task-organizes the battalion, fitting the forces to the ground, mission, and enemy situation. Divisions provide CS and CSS force multipliers. Corps conducts operational-level warfare, providing additional CS and CSS assets in accordance with the corps main effort.

(3) To execute doctrine, the infantry battalions require adequate troop strength; an organic antiarmor capability; supporting arms; optimized task organization based on the mission; and adequate support. These requirements are met through the organization of the infantry battalions and through augmentation and task organization where required.

b. Types of Battalions. There are six basic types of infantry battalions: mechanized infantry, air assault, airborne, ranger, light, and mountain. The fundamental combat mission of the infantry battalion, regardless of type, is to destroy or capture the enemy by means of fire and maneuver; or to repel his assault by fire, close combat, and counterattack. To accomplish specific missions, the battalion is normally augmented with combat, CS, and CSS assets.

c. Task Organization. Normally, infantry battalions operate as table of organization and equipment (TOE) units only in garrison. For training and for combat, they are task organized for the mission at hand. Task organizing tailors the unit to get the most from its capabilities and to minimize its limitations. It is a temporary grouping of forces designed to accomplish a particular mission. Task organization involves the distribution of available assets to subordinate control headquarters by attaching or placing assets under OPCON to the subordinate. Task organization is made after analysis of the mission and all of the other METT-TC factors. When developing the task organization, the commander must clearly understand the capabilities and limitations of his organic and supporting units; he must consider the existing C2 relationships.

1-29. Organization of the Mechanized Infantry and Armored Battalions

Mechanized infantry battalions and armored battalions are organized, equipped, and trained to accomplish specific missions; each type battalion has unique capabilities and limitations.

a. Missions.

(1) The mission of the mechanized infantry battalion is to destroy or capture the enemy by means of fire and maneuver, or to repel his assault by fire, close combat, and counterattack.

(2) The mission of the armored battalion is to close with and destroy enemy forces using fire, maneuver, and shock effect, or to repel his assault by fire and counterattack.

(3) Battalion TF accomplish missions and tasks as part of a brigade's operation. Occasionally, TF will conduct operations directly under a division's or an ACR's control; such as participating in the higher headquarters covering force; acting as a reserve; or forming a TCF in rear area operations.

b. Capabilities.

(1) The capability of the mechanized infantry battalion and the armored battalion is increased through task organization. Based on situational estimates, the brigade commander task-organizes armored and mechanized infantry battalions by cross-attaching companies between these units. As a rule, cross-attachment is done at battalion level because it has the necessary C2 and support capabilities to employ combined arms formations. The brigade commander determines the mix of companies in a TF. Similarly, the TF commander may cross-attach platoons to form company teams for specific missions.

(2) Tank and mechanized infantry battalion TF apply their mobility, firepower, and shock effect to—

- Conduct sustained combat operations in all environments.
- Accomplish rapid movement and limited penetrations.
- Exploit success and pursue a defeated enemy as part of a larger formation.
- Conduct security operations (advance, flank, or rear guard) for a larger force.
- Conduct defensive, retrograde, or other operations over assigned areas.
- Conduct offensive operations.

c. *Limitations.*

(1) Because of the high density of tracked vehicles, the battalion has the following limitations:

- Mobility and firepower are restricted by urban areas, dense jungles and forests, very steep and rugged terrain, and significant water obstacles.
- Strategic mobility is limited by substantial quantities of heavy equipment.
- Consumption of supply items is high, especially Classes III, V, and IX.

(2) Battalions are task-organized according to mission; they are routinely augmented to improve engineer, fire support, air defense, intelligence, and CSS capabilities.

1-30. Battalion Task Force on the AirLand Battlefield

a. The foundation of AirLand Battle doctrine at the TF level is classical maneuver warfare. In its simplest form, maneuver warfare involves using a part of the force to find, then contain the enemy, while the remainder of the force moves to a position of advantage and attacks his weakest point—usually a flank or the rear. The goal is to mass enough combat power at the critical place and time to destroy or threaten the enemy with destruction, while preserving freedom for future action.

b. The TF commander must understand the intent of the brigade and division commander to properly employ his force. The TF commander develops his intent and concept and accepts risks to achieve decisive results. He seizes the initiative early and conducts offensive action aimed at imposing his will on the enemy. The objective of his maneuver is to position strength against weakness, throw the enemy off balance, and aggressively follow-up to defeat and destroy the enemy.

Section VII. THE BATTALION STAFF

1-31. Command and Control Responsibilities of the Battalion

The commander establishes a standard C2 system by defining the functions of key individuals, organizations, and facilities. He organizes his staff in a manner to accomplish the mission. He will develop a basic organization flexible enough to be modified to meet changing situations. This section discusses the individual and staff functions and responsibilities and how they are organized to facilitate C2.

1-32. Staff

a. Commander. The commander commands and controls subordinate combat, CS, and CSS elements that are organic or attached to his unit or under its OPCON. The commander's main concerns are to accomplish his unit's mission and to ensure the welfare of his soldiers.

(1) The commander cannot win the battle alone. He must rely on his staff and subordinate commanders for advice and assistance in planning and supervising operations. He must completely understand their limits and capabilities. He must train subordinate commanders to execute his concept in his absence. Also, he must cross-train his staff to continue unit operation when staff elements suffer combat losses.

(2) The staff reduces the demands on the commander's time; they assist him by—

- Providing information.
- Making estimates and recommendations.
- Preparing plans and orders.
- Supervising the execution of orders issued by, or in the name of, the commander.

The commander assigns clear-cut responsibility for functions to unit staff officers to ensure that conflicts do not arise. As a rule, staff officers are delegated the authority to say "yes" to requests by subordinate unit commanders. They defer to the commander when the answer is "no." The staff must be responsive to subordinate unit commanders.

b. Executive Officer. The executive officer (XO) is second in command and is the principal assistant to the battalion commander. The XO is prepared to assume the duties of the commander. He formulates and announces staff operating policies and ensures the commander and staff are informed on matters affecting the command. The XO ensures that—

- The required liaison is established.

- All staff officers, unless otherwise instructed by the commander, inform him (the XO) of any recommendations or information they gave directly to the commander, or of any instructions they received directly from the commander.

- The XO functions as the *Chief of Staff* and directs and coordinates staff activities.

He represents the commander, when required, and exercises supervision of the tactical operations center (TOC) and its operations.

c. Command Sergeant Major. The CSM is the senior noncommissioned officer (NCO) in the unit. He acts in the name of the commander when dealing with other NCO in the unit; he is the commander's primary advisor concerning the enlisted ranks. He should understand the administrative, logistical, and operational functions of the unit to which he is assigned. Since he is the senior enlisted soldier in the unit, his attention should be focused on operations, training, and how well the commander's decisions and policies are being carried out. He is the senior enlisted soldier who is responsible to the commander for training in the organization. He coaches and trains first sergeants (1SGT) and platoon sergeants (SGT); he works very closely with company commanders in this regard. He maintains close contact with subordinate and attached unit NCOs. The CSM may act as the commander's representative in supervising critical aspects of an operation. The CSM can lead the advance/quartering party during a major movement. He may also help in the CSS effort during the battle.

1-33. Coordinating Staff

a. Adjutant (US Army).

(1) The Adjutant (US Army) (S1) has primary responsibility to provide all activities and functions associated to sustain personnel manning of the unit, provide personnel support, and other human resource support to service members. The S1 section is principally structured to provide critical wartime functions of personnel strength accounting and casualty reporting. Battalion S1 sections can provide limited personnel services and support within the battalion, but require augmentation support to provide the full spectrum of human resource support and services. The S1 participates in the planning process, personnel estimates, loss estimates, and recommends replacement priorities. Within the battalion, the S1 coordinates EPW support, limited legal support, CHS planning (in conjunction with the battalion surgeon), and public affairs. The battalion S1 also coordinates legal actions with the brigade S1.

(2) Information linkages and split-base operating procedures should minimize the distinction between wartime/peacetime and forward/rear in regards to personnel support. Personnel (S1) sections provide additional services at the home station. When deployed, the S1 performs replacement operations, postal operations, strength management, personnel actions (awards, promotions, evaluations, reassignments, and military pay), morale welfare and recreation support, and Red Cross coordination. The commander must leverage his reach capability with these functions.

(3) During deployment, the S1 operates from the combat trains and is normally collocated with the Logistics Officer (US Army) (S4). While collocated, functional responsibilities between the S1 and S4 are not intended to be interchangeable.

b. Intelligence Officer (US Army). The Intelligence Officer (US Army) (S2) exercises overall staff responsibility for intelligence. He conducts the IPB with the commander and Operations and Training Officer (US Army) (S3) using—

- Higher collection sources.
- Ground and aerial RECON.
- Observation posts.
- Ground surveillance radar.
- Target acquisition.
- Electronic warfare assets.
- Organic scout platoon.

In conjunction with the IPB process, he prepares and disseminates intelligence estimates. The tactical intelligence officer assists the S2. The tactical intelligence officer is part of a two-man battalion information coordination center (BICC). The BICC's primary responsibility is to manage the unit's intelligence collecting, processing, and disseminating effort for the S2. The BICC develops and initiates the reconnaissance and surveillance (R&S) plan; identifies requirements that cannot be met by the battalion's assets; and notifies the brigade S2.

c. Operations and Training Officer (US Army). The S3, as the operations officer, is the commander's principal assistant for coordinating and planning the battle. The S3—

- Monitors the battle.
- Makes sure that CS assets are provided when and where required.
- Anticipates developing situations.

He advises the commander on—

- Courses of action (COA).
- Combat and CS matters.
- Organization and training.
- Operational matters during the battle.

He prepares the operations estimate and conducts planning and coordination with other staff sections resulting in published operation orders (OPORD), operation plans (OPLAN), and training programs.

In conjunction with his planning duties, he is responsible for PSYOP; EW activities, OPSEC; deception; and, in conjunction with the S4, tactical troop movement. He establishes priorities for communications to support tactical operations and coordinates with the XO and the battalion signal officer on the location of the main CP.

(1) Operations and Training Officer (US Army) (Air). The S3 (Air), the principal assistant to the S3, is normally in the TOC. He assumes the duties of the S3 in his absence. He coordinates the employment of close air support with the fire support element (FSE) and the TACP, as well as the air defense section leader.

(2) Nuclear, biological, and chemical personnel. The assistant S3/chemical officer is assigned to the S3 section of combat battalions with a chemical NCO as his assistant. A decontamination specialist is assigned to the HHC of airborne and air assault battalions. The chemical officer and NCO train and supervise the battalion decontamination crew. During combat operations, chemical personnel provide a 24-hour capability within the S3 section to receive, correlate, and disseminate information on NBC attacks. They consolidate subordinate units' operational exposure guide (OEG) radiation status and report to higher headquarters as required. They provide recommendations concerning mission-oriented protective posture (MOPP) levels and employment of supporting NBC RECON and smoke units. If the unit comes under NBC attack, battalion NBC personnel organize and establish a battalion NBC center. They supervise activities of the radiological survey and monitoring teams and the chemical detection teams. The NBC personnel also coordinate and supervise decontamination missions (less patient decontamination) conducted with or without support-level decontamination assets.

d. Logistics Officer (US Army). The S4 has primary staff responsibility for determining CSS requirements and priorities. The S4 is responsible for CSS planning in the military decision-making process. His section is responsible for the procurement, receipt, storage, and distribution of supplies; and for transportation of units, personnel, and CSS items to their required locations. He designates lines of communications (LOC) and supply routes and locations of CSS elements; and prepares and develops CSS plans in concert with the current tactical plan. The S4 is responsible for the preparation, authentication, and distribution of CSS support plans and orders when published separately. The S4 establishes the requirements for civilian labor and the collection and disposal of excess property, salvage, and captured material.

e. Command and Control, Communications, and Computer (C4) Operations Officer (US Army). The Assistant Chief of Staff, (Signal) (G6)/Communications Staff Officer (US Army) (S6), C4 operations officer, is the principal staff officer for all matters concerning C4. A G6/S6 is located at all echelons of command from battalion through corps. He is responsible for advising the commander, staff, and subordinate commanders on C4 operational matters. Command and control, communications, and computer operations are inclusive of network operations and information management. Network operations include network management, information dissemination management, and information assurance. Information management includes relevant information and information systems functions.

f. Battalion Maintenance Officer. The battalion maintenance officer (BMO) plans, coordinates, and supervises the maintenance and recovery efforts of the maintenance platoon and ensures that adequate maintenance support is provided to the TF. Although he is a staff officer in the battalion headquarters, he is also the maintenance platoon leader. The maintenance warrant officer assists the BMO by providing technical assistance and supervision of the maintenance platoon. The BMO supervises the unit maintenance collection point in the armored and mechanized infantry battalions only.

1-34. Special Staff

a. Battalion Surgeon. In this role the battalion surgeon is a special staff officer and advisor to the battalion commander on employment of the medical platoon and on the health of the battalion. He is also the supervising physician (medical officer/field surgeon) of the medical platoon's treatment squad. This officer is responsible for all CHS provided by the platoon. His responsibilities include—

- Planning and directing CHS for the maneuver battalion. He does this in conjunction with the battalion S1, who is the coordinating staff officer responsible to the commander for health and welfare of the troops.
- Advising the maneuver battalion commander and his staff on CHS operations and the medical threat.
- Supervising the health, welfare, organizational training, administration, discipline, maintenance of equipment, supply functions, and employment of medical platoon personnel.
- Examining, diagnosing, treating, and prescribing courses of treatment for patients, to include ATM.
- Training combat lifesavers (CLS).
- Supervising the battalion CSC program.
- Supporting humanitarian assistance programs, when directed.
- Providing PVNTMED support for the battalion. Requesting PVNTMED support from the brigade for PVNTMED requirements beyond his (battalion surgeon) capabilities.
- Planning and overseeing PVNTMED training for battalion personnel.
- Advising the commander on the health of the battalion.
- Supervising the training of unit field sanitation teams.
- Ensuring the field health records are maintained according to AR 40-66.
- Advising the commander on the effects of the Geneva Conventions on CHS. See Appendix A, The Geneva Conventions, for additional information.

b. Chaplain. The chaplain advises the commander and staff on religious matters and of the influence of indigenous religious groups and customs on the commander's courses of actions. The chaplain is a member of the battalion's unit ministry team (UMT). This team is the staff section that provides religious support to the battalion. The team advises the commander on unit morale and ethical issues and assists in meeting the religious and spiritual needs of the soldiers. The team consists of a chaplain and a chaplain assistant. The chaplain provides the clergy-related support to the unit. These include worship and prayer services, funeral and memorial services, and in-depth grief counseling. The chaplain assistant provides the administrative and logistical management for the team as well as the team's security. The chaplain exercises the necessary staff authority for developing, coordinating, and executing the religious support plan. Additionally, the chaplain facilitates soldiers' free exercise of their religious rights, beliefs, and worship practices and makes recommendations for ethical decision-making and moral leadership programs. See FM 16-1 for definitive information on religious support for the battalion.

1-35. Other Staff Assets

a. Headquarters and Headquarters Company Commander. The HHC commander has the responsibility of ensuring that the command facilities are provided logistical support. Normally, he places his XO with the main CP to supervise support, security, and movement. The HHC commander locates himself at the field trains CP to monitor and coordinate all battalion activities there. He uses landlines and messengers to control all elements in the field trains and communicates with the combat trains using the administration/logistics net (a frequency modulated [FM] radio net). The HHC commander is available for other tactical missions as dictated by the estimate of the situation. These roles normally come into play during operations other than sustained ground combat. They may include coordination and control of the RECON/counter-RECON effort; combat patrols; or any other task designated by the battalion commander.

b. Physician Assistant. The physician assistant (PA), normally a Captain, Army Medical Specialist Corps, performs general technical health care and administrative duties. The PA is ATM-qualified and works under the clinical supervision of the medical officer. This officer serves as the medical platoon leader in the absence of an assigned physician.

c. Field Medical Assistant. The field medical assistant, a Medical Service Corps (MS) officer, normally a Lieutenant, is the operations/readiness officer for the medical platoon. He is the principal assistant to the battalion surgeon for operations, administration, and logistics. The field medical assistant coordinates CHS operations with the battalion S3 and S4 and coordinates patient evacuation with the FSMC.

d. Fire Support Officer. The integration of fire support into the maneuver operation is a decisive factor in the success of battle. The maneuver commander is responsible for the whole of his operation including the fire support plan. The fire support officer (FSO) is responsible for advising the commander on the best available fire support resources; for developing the fire support plan; for issuing the necessary orders in the name of the commander; and for implementing the approved fire support plan. The FSO normally locates with the commander, but it may be necessary to locate where he can communicate best.

e. Air Defense Artillery Officer. The senior leader of any supporting ADA unit(s) advises the commander on the employment of ADA assets. During the planning process, he is at the TOC to ensure the integration of air defense into the concept of operation. During the execution of the plan, he positions himself in a location that will enable him to best C2 the air defense assets. He monitors the command net to remain responsive to the needs of the commander. He also monitors the early warning net to assist in the acquisition and dissemination of early warning information as a member of the Army airspace C2 system.

f. Engineers. The leader of the supporting engineer unit advises the commander on the employment of engineer assets. During the initial planning, he is at the TOC to advise the commander on employment of his unit. During the battle, the engineer unit provides a representative with a radio at the TOC, if possible, to coordinate the engineer effort. If no representative is available, the TOC periodically monitors the engineer net. Regardless of the system used, the engineer leader is responsible for maintaining constant communications with the battalion.

g. Antiarmor Company Commander/Platoon Leader (Light Battalion). This leader advises the commander on the tactical employment of his weapon systems. He may serve as a fourth maneuver element or as an alternate battalion CP when properly task-organized. The first alternate CP for the battalion is the combat trains CP (CTCP).

h. Scout Platoon Leader. He advises the commander and the S2 on the employment of his element. He is responsible for conducting tactical RECON in support of the battalion. He assists the S2 in developing the R&S plan.

i. Battalion Mortar Platoon Leader. He advises the battalion commander and the FSO on tactical employment of the battalion mortar platoon; he may assist the FSO with his fire support coordinator responsibilities. His platoon headquarters may also serve as an alternate CP.

CHAPTER 2

DIVISION COMBAT HEALTH SUPPORT

Section I. OVERVIEW OF DIVISION AND CORPS MEDICAL ASSETS

2-1. Tables of Organization and Equipment

The TOE provides a model for fielding a unit at full capability, or at a reduced capability if resource constraints so mandate. The TOE also specifies the capabilities that the unit has to accomplish its mission. Tables of organizations and equipment are scheduled for revision when changes in doctrine occur, upon introduction of new or improved equipment, or to incorporate more effective organizational design. New TOEs are developed to accommodate the requirements of new organizations. If the TOE is not scheduled for revision or replacement by a new TOE, it will be scheduled for cyclic review every 3 years. There are different editions of TOE that are identified as under the TOE modernization system. For example, A-Edition TOE were developed to identify restructuring initiatives for existing L-Edition TOE. Example of TOE editions based on modernization of the Army include—

- Division 86, H-Edition TOE.
- Army of Excellence (AirLand Battle/Medical Force 2000), L-Edition TOE.
- Medical Reengineering Initiative (MRI), A-Edition TOE.
- Force XXI (Digitized Division), F-Edition TOE.

The MTOE is the document that is seen at the unit level. The MTOE is a modified version of a Headquarters, Department of the Army (DA)-approved TOE that prescribes the unit organization, personnel, and equipment necessary to perform a mission in a specific geographical or operational environment. At unit level, the MTOE is the base document for requesting personnel and equipment; distributing personnel and equipment resources; unit status reporting; and reporting supply and maintenance status. In addition to the TOE, some organizations have a table of distribution and allowances (TDA). The TDA prescribes the organizational structure for a unit having a support mission for which a TOE does not exist and which may include civilian positions. For the remainder of this chapter only the Army of Excellence/AirLand Battle and the Force XXI/Digitized Division TOE (with some information pertaining to MRI units) will be discussed.

2-2. Division Medical Assets, L-Edition TOE (Army of Excellence/AirLand Battle)

The Army of Excellence/L-Edition TOEs were developed to support the AirLand Battle/Medical Force 2000 doctrine. They will be in the Army's inventory until the Army completes its Army XXI and other modernization initiatives. Medical assets that are organic to the division with units under the L-Edition TOE include—

- Division surgeon's section (DSS) that has a surgeon and three additional medical personnel.
- Division support command, division medical operations center (DMOC), which could have up to 18 personnel assigned, depending on type of communications equipment the DMOC is using.

- Main support battalion, health service support officer (HSSO).
- Main support medical company.
    - Headquarters section.
    - Treatment platoon.
    - Ambulance platoon.
    - Division medical supply office.
    - Mental health section.
    - Preventive medicine section.
    - Optometry section.
- Forward support battalion, HSSO.
- Forward support medical company.
    - Headquarters section.
    - Treatment platoon.
    - Ambulance platoon.
- Aviation brigade flight surgeon's section

In addition, medical platoons/section are assigned to the following combat arms battalions or squadrons:

- Armored.
- Mechanized infantry.
- Airborne infantry.
- Air assault.
- Light infantry.
- Armored cavalry.
- Artillery.

- Reconnaissance, surveillance, and target acquisition (RSTA) squadron.
- Air defense artillery battalion.

Also, medical personnel may be assigned to the following units.

- Engineer battalion.
- Signal battalion.
- Antitank company.
- Military police company.

2-3. Division Medical Assets, F-Edition TOE (Force XXI/Digitized Division)

The F-Edition TOEs were developed to support the Force XXI/Division Redesign Initiative and Force XXI doctrine. Digitization of the new Force XXI Division is a high priority.

NOTE

Digitization is defined as the application of information to acquire, exchange, and employ timely battlefield information. It will *enhance situational understanding* and provide the means for information dominance by enabling friendly forces (ten divisions, Reserve Components, and joint/combined forces) to *share a common picture of the battlefield* while communicating and targeting in real or near real-time. Digitization will reduce the "fog of war" and *decrease decision-making time* by *optimizing the flow of information*. It will allow the "orchestration" of combat power at critical times and places faster than an adversary can. It will contribute *increased lethality, survivability, and operational tempo* while *reducing the potential for fratricide* to ensure *seamless digital communications* from the sustaining base to the tactical and strategic levels.

Medical assets that are organic to the division with units under the F-Edition TOE include—

- Division surgeon's section (a surgeon and 12 additional medical personnel).
- Division support command medical operations branch, including an officer and an NCO.
- Division support command medical materiel management branch (MMMB) (one officer and one NCO).

- Division support battalion HSSO.
- Division support medical company (DSMC).
  - Headquarters section.
  - Treatment platoon.
  - Ambulance platoon.
  - Mental health section (a division psychiatrist, an NCO, and one MH specialist are assigned to this section.
  - Preventive medicine section (one environmental science officer and a PVNTMED NCO are assigned to this section).
  - Optometry section.
- Forward support battalion, CHS cell/HSSO (an officer and an NCO are assigned to this section).
- Forward support medical company.
  - Headquarters section.
  - Treatment platoon (one less treatment squad than the L-Edition TOE).
  - Ambulance platoon.
  - Mental health section (a division psychiatrist, an NCO, and one MH specialist are assigned this section).
  - Preventive medicine section (one environmental science officer and a PVNTMED NCO are assigned to this section).
- Maneuver brigade surgeon's section (BSS) (six personnel assigned).
- Medical platoons/sections are assigned to the following combat arms and CS battalions:
  - Armored.
  - Mechanized infantry.
  - Artillery.

- Engineer.
- Signal.

- Medical personnel may be assigned to the following units:
  - Engineer company.
  - Antitank company.

2-4. Division Surgeon, L-Edition TOE (Army of Excellence/AirLand Battle)

The division surgeon is an Medical Corps (MC) officer, area of concentration (AOC) 60A. He is a special staff officer and normally coordinates his CHS activities through the G1. Generally, the surgeon's duties are administrative; the division commander charges him with full responsibility for the technical control of all medical activities in the command. The division surgeon's staff is assigned to the DSS of the division HHC. Personnel assigned to this section include an operations NCO (MOS 91W40), a clerk typist (MOS 71L10), and a patient administration specialist (MOS 71G10). These personnel along with the DMOC staff located in the DISCOM, assist the division surgeon in the performance of his duties. The division surgeon's responsibilities include—

- Advising on the health status of the command and of the occupied or friendly territory within the commander's area of responsibility.

- Briefing the division commander and/or his representative during all routine and emergency division briefings on CHS operations.

- Participating in the preparation of division OPLAN and contingency plans and identifying potential medical hazards associated with geographical locations and climatic conditions.

- Determining reporting requirements and frequencies.

- Advising on the health effects of the environment.

- Advising on the health effects of NBC devices/weapons, to include OEG.

- Exercising technical supervision of subordinate brigade surgeons, physicians, and PAs.

- Providing consultation and mentoring to subordinate brigade surgeons, physicians, and PAs.

- Advising on the health effects of directed-energy devices/weapons.

- Determining requirements for the requisition, procurement, storage, maintenance, distribution management, and documentation of Class VIII supplies within the division.

- Monitoring critical Class VIII items and keeping the division commander informed.

- Determining requirements for medical personnel and making recommendations concerning their assignments.

- Coordinating with medical unit commanders, to include leaders of medical platoons and sections, for continuous CHS.

- Submitting to higher headquarters those recommendations on professional medical problems that require research and development.

- Recommending use of captured medical supplies in support of EPW and other recipients.

- Advising on medical intelligence requirements (including the examination and processing of captured medical supplies as directed by the corps surgeon).

- Providing recommendations on allocation and redistribution of AMEDD personnel, CHL, and CHS during the reconstitution process.

- Advising commanders about the PVNTMED aspects of reconstitution and availability and use of CSC teams.

- Forwarding the Command Health Report (RCS MED-3 [R7]) according to Chapter 3, AR 40-5.

- Advising commanders on the effects of accumulated fatigue, radiation exposure, possible delayed effects from exposure to chemical warfare (CW) or biological warfare (BW) agents, and use of countermeasures and pretreatments.

- Advising commanders on disposition of personnel exposed to lethal, but not immediately life-threatening, doses of radiation or CW and BW agents.

- Ensuring the division's CHS annex is developed for all contingency plans. For CHS planning factors, see FM 8-55.

- Initiating, through the division commander and the G3, medical training, first aid training, and CLS training programs for the division.

- Overseeing the continuing health education program for the division and ensuring compliance with AR 351-3.

- Ensuring that all physicians and nonphysician health care providers have gone through a credentialing committee according to AR 40-68 to validate clinical privileges.

- Ensuring that clear and accurate patient records are maintained of all clinical encounters for supported/deployed personnel through the use of appropriate forms as directed by AR 40-66. See Appendix B for management of individual health records in the field. For additional information on the division surgeon, see FMs 8-10-3 and 8-10-5.

2-5. Division Surgeon, F-Edition TOE (Force XXI/Digitized Division)

The division surgeon, an MC officer (Lieutenant Colonel [LTC], AOC 60A00), is a division-level special staff officer. He normally works under the staff supervision of the division Chief of Staff. The division surgeon is responsible for the technical control of all medical activities in the command. He oversees and coordinates CHS activities through the DSS. The division surgeon advises the division commander on all medical or medical-related issues. The division surgeon's responsibilities are the same as those identified above except for—

- Briefing the division commander and/or his representative during all routine and emergency division briefings on CHS operations. This is normally accomplished using the Combat Service Support Control System (CSSCS).

- Monitoring the status of critical Class VIII items list and providing the G4 a list of medical items that should be a part of the commander's tracked items list.

- Developing the division's CHS annex for all contingency plans. For CHS planning factors, see FM 8-55.

- Ensuring that clear and accurate patient records are maintained of all clinical encounters for supported/deployed personnel through the use of appropriate forms as directed by AR 40-66. See Appendix B for management of individual health records in the field. Also, digital patient records at the division and brigade level will be available through the fielding of Medical Communications for Combat Casualty Care (MC4) and the Theater Medical Information Program (TMIP); see Chapter 3. For additional information on the division surgeon, see FM 4-02.21.

2-6. Division Surgeon's Section, F-Edition TOE (Force XXI/Digitized Division)

The DSS is normally located with the division main and consists of a medical plans and operations cell, a CHL cell, a patient disposition and reports cell, and a PVNTMED cell. Figure 2-1 shows the typical organization and staffing of the DSS.

a. Medical Plans and Operations Cell. The medical plans and operations cell is responsible for coordinating, planning, synchronizing, rehearsing, and conducting CHS for the division. For definitive information on the organization, functions, and operations of this section, see FM 4-02.21.

b. Combat Health Logistics Cell. The CHL cell is responsible for planning, coordinating, and prioritizing CHL and medical equipment maintenance programs for the division. The CHL cell is staffed with a health service materiel officer (HSMO). The HSMO (Major, AOC 70K67) works closely with the DISCOM MMMB and medical logistics (MEDLOG) company. The HSMO coordinates and oversees the CHL support for the division.

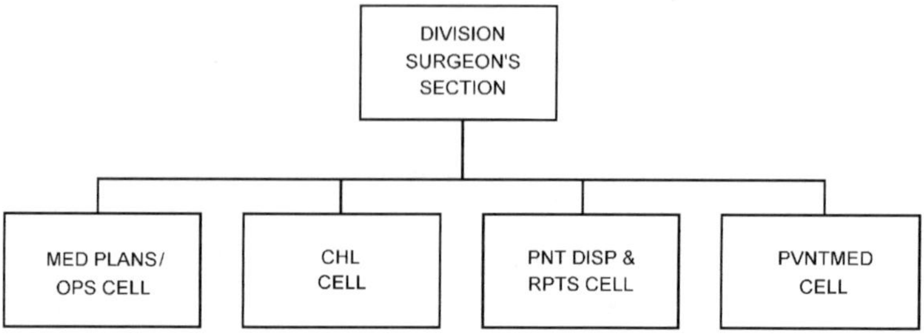

Figure 2-1. Division surgeon's section.

c. Patient Disposition and Reports Cell. The patient disposition and reports cell is responsible for coordinating patient disposition throughout the division. The branch obtains and coordinates disposition of patients with the medical plans and operations cell and the corps medical regulating office(r) (MRO). It prepares and forwards appropriate medical statistical reports as required. The patient disposition and reports cell is staffed with a patient administration NCO and two patient administration specialists.

d. Preventive Medicine Cell. The division PVNTMED cell is responsible for—

- Supervising the command PVNTMED program, to include health assessment and medical surveillance; see AR 40-5 and FM 4-02.17.

- Ensuring PVNTMED measures are implemented that protect division personnel against food-, water-, and vectorborne diseases, as well as environmental injuries (for example, heat and cold injuries).

- Monitor disease trends within the division.

The PVNTMED missions are accomplished according to the division CHS plan and coordinated by the PVNTMED officer through the medical plans and operations cell with the DSMC and FSMCs. Division PVNTMED personnel provide advice and consultation in the areas of environmental sanitation, epidemiology, and entomology, as well as limited sanitary engineering services and pest management. Additional information pertaining to the PVNTMED personnel and their specific functions is discussed in FMs 4-02.17, 8-10, 8-10-1, and 8-10-3. The PVNTMED cell is staffed with a PVNTMED officer. The PVNTMED officer (Major, AOC 60C00) is responsible for the implementation of the command PVNTMED program.

2-7. Corps Medical Assets in Support of Divisions

Corps medical units in GS and DS of the division are normally assigned to the corps medical command (MEDCOM) or medical brigade. The MEDCOM/brigade will provide subordinate units to support the division by establishing a command relationship of OPCON or attachment. The MEDCOM/brigade could also choose to maintain only a support relationship of DS or GS to support the division. The division surgeon and DSS (Force XXI) and the DMOC under Army of Excellence interface with corps medical units according to the MEDCOM/brigade tactical standing operating procedures (TSOP). The DSS or DMOC and other division staff elements must be prepared to integrate corps-level medical units/elements into the medical, as well as the logistical, support structure. The MEDCOM/brigade will normally deploy a liaison officer to the division to coordinate and synchronize corps CHS. Information concerning the organization, functions, and responsibilities of the corps MEDCOM/brigade is found in FM 8-10.

a. Corps Medical Command and Medical Brigade. The corps MEDCOM and medical brigade provide C2, including—

- Staff planning.
- Supervision of operations.
- Administration of the assigned and attached units.

b. Medical Logistics Battalion. The MEDLOG battalion is organic to the corps medical brigade. The MEDLOG battalion provides C2 for assigned MEDLOG companies and the blood support detachment. The MEDLOG battalion is responsible for receiving, storing, and distributing medical materiel; single and multivision optical fabrication and repair; medical maintenance; blood and blood product collection, manufacturing, and distribution; medical gas production and distribution; and building of medical assemblages/push packages. The MEDLOG battalion will employ standard state-of-the-art MEDLOG information management and communications systems, to include satellite links.

(1) *Medical logistics company.* The MEDLOG company provides Class VIII supplies, DS/GS medical maintenance, and optical support. The MEDLOG company will use line-item requisitioning to support customers and will have the capabilities of building and maintaining preconfigured push packages in support of forward deployed medical units.

(2) *Blood support detachment.* The MEDLOG battalion's blood support detachment serves as the Army's blood supply unit (BSU). Blood and blood products will be stored and distributed under rigid specifications and managed by standard automated systems. Air movement will be the mode of choice for transporting blood and blood products. Army blood support in the AO will be the responsibility of the supporting MEDLOG battalion. The MEDLOG battalion's blood support detachment will collect, manufacture, receive, store, and distribute blood and blood products on an area basis.

c. Medical Evacuation Battalion. The headquarters and headquarters detachment, MEDEVAC battalion serves as the central manager of ground and air evacuation assets in the corps. Its mission is to provide C2 of ground and air MEDEVAC units within its AO. Information pertaining to the organization, functions, and capabilities of this unit is discussed in FM 8-10-6; air ambulance operations in support of the division are discussed in FMs 4-02.21, 8-10-3, and 8-10-26. A ground ambulance and one air ambulance company are normally placed in support of a division. The air ambulance company provides aeromedical evacuation on a DS basis. This company may be attached for support (less OPCON) to the division aviation brigade. Air ambulances will operate from the DSA and BSA, providing 24-hour immediate response MEDEVAC capability.

d. Medical Detachment, Combat Stress Control. The CSC detachment provides DS to a division's maneuver brigades and GS/reinforcing support to the DSA, including corps units in those areas. The detachment must function with its elements widely dispersed, some working in and for the supported division and others working in the corps. For definitive guidance on the medical detachment, CSC, see FM 8-51.

e. Veterinary Elements. The division will have corps veterinary team/personnel positioned at Class I supply points. They may also be deployed to the division to monitor and evaluate zoonotic diseases and environmental data, to include food exposed to NBC agents. Veterinary personnel will provide animal care for military working dogs and may perform investigations and postmortem examination of reported animal deaths. For additional information see FM 8-10-18.

f. Preventive Medicine Elements. Corps PVNTMED team/personnel may be deployed into the division when requested to augment division PVNTMED sections. Corps PVNTMED personnel provide advice and consultation in the areas of environmental sanitation, epidemiology, and entomology, as well as sanitary engineering services and pest management. Additional information pertaining to corps PVNTMED units and their specific functions is discussed in FM 4-02.17.

g. Dental Services. In planning the concept of operations, unit-level dental support is dependent upon corps-level area dental support assets in numbers sufficient enough to support the manpower requirement criteria for operational dental care. Unit dental support relies on corps-level area dental support units for assistance in providing operational care. Modules of area dental support units also augment or reconstitute unit dental elements when required. Corps-level dental units provide dental modules to reinforce or reconstitute the division dental modules when necessary and to operate field dental clinics. See FM 4-02.19 for definitive information on dental services.

Section II. DIVISION SUPPORT COMMAND MEDICAL UNITS AND ELEMENTS

2-8. Division Support Command, L-Edition TOE (Army of Excellence/AirLand Battle)

The DISCOM is organized to provide the maximum amount of CSS within prescribed strength limitations while providing the most effective and responsive support to tactical units in a combat environment. In order to provide responsive support to the tactical commander, logistics, medical, and personnel services support must be effectively organized and positioned as far forward as necessary to support the tactical plan. Division-level CHS for the Army of Excellence division is coordinated and provided by the DISCOM medical elements listed below:

- Division medical operations center, DISCOM HHC, located in the DSA.

- Main support battalion.
- Main support medical company, MSB, located in the DSA.
- Forward support battalion.
- Forward support medical company, FSB, located in the BSA.

a. Division Medical Operations Center. The DMOC's mission is to plan, coordinate, and synchronize the division's CHS with technical medical advice from the division surgeon. The division surgeon and the DMOC chief have joint responsibilities for CHS operations in the division. Their staff positions in the division and DISCOM require a close working relationship and coordination of their CHS activities. The DMOC staff is responsible to the DISCOM commander for staff supervision of CHS within the DISCOM. The division surgeon and DMOC chief will develop operating procedures that will enhance the flow of information and facilitate the synchronization of CHS operations within the division. It is imperative that the division surgeon and the DMOC chief work as a team. Both share equal responsibility for planning and overseeing CHS operations. The DMOC is responsible for monitoring CHS activities within the division area and keeping the DISCOM commander informed of the status of CHS. The division surgeon is informed of the DISCOM's CHS status through reports prescribed by the TSOP. For definitive information on the DMOC, see FMs 8-10-3 and 63-2.

b. Main Support Battalion. Information pertaining to the structure and operations of the MSB is provided in FM 63-21. The DMOC will interface with elements of the MSB, as required and approved by the DISCOM commander. The DMOC may interface with elements of the MSB through the DISCOM support operations section. The interactions and coordination between the DMOC and the MSB are driven by CHS requirements of the division and changes with the tactical situation. The chief, DMOC, and the MSB commander must develop policies and procedures that clearly delineate responsibilities and coordination requirements for an effective working relationship. Tasking of the MSMC elements by the DISCOM will be through command channels.

c. Main Support Medical Company. The MSMC provides division- and unit-level CHS and medical staff advice and assistance on an area basis to units operating in the DSA. Combat health support operations are coordinated by the DISCOM DMOC medical operations branch through technical channels. The DISCOM will task elements of the MSMC through command channels to provide division-level CHS. The interface between the MSMC and the DMOC is essential for providing required division CHS. For definitive information of the MSMC, see FMs 8-10-1, 8-10-3, and 63-21.

d. Forward Support Battalion. The DMOC will interface with elements of the FSB as required and approved by the DISCOM commander. The DMOC may interface with elements of the FSB through the DISCOM support operations section. This interface between the DMOC and elements of the FSB is driven by CHS requirements in the forward areas. This information will assist the DMOC in planning, coordinating, and managing division medical elements and resources in support of the battle. Communications and coordination between elements of the DMOC and the FSB are essential for successful accomplishment of the DMOC's and FSB's CHS mission. For additional information, see FM 63-20.

e. Forward Support Medical Company. The FSMC provides CHS for the brigade as well as area medical support for the brigade rear. Combat health support operations are coordinated by the FSMC commander and the FSB HSSO. The DISCOM tasks elements of the FSMC through command channels to provide division-level CHS. The FSMC commander positions are documented 05A, AMEDD immaterial, meaning any qualified AMEDD officer can assume command. When the FSMC commander is not a physician, medical decisions and technical supervision of physicians is performed by the senior physician assigned to the FSMC. The FSMC commander keeps the FSB commander informed on the CHS aspect of FSB operations and the health of the command. He regularly attends FSB staff meetings to obtain information to facilitate the execution of medical operations. He provides staff estimates and assists the FSB and brigade staffs with development of the CHS plan. For additional information on the FSMC commander, see FMs 8-10-1 and 63-20.

2-9. Division Support Command, F-Edition TOE (Force XXI/Digitized Division)

The DISCOM is a multifunctional organization capable of providing, coordinating, and synchronizing logistical support to the division. The DISCOM's mission of sustaining the divisions combat power is more critical than ever. The DISCOM consists of an FSB, a division support battalion (DSB), a division aviation support battalion, and the HHC. The DISCOM provides CSS for the division. It provides arming through its Class V operations, fueling through Class III operations, repairing through its maintenance operations, transportation through the truck company and the supply and transportation sections in the FSB, and sustaining, through the provision of rations, individual equipment and medical support. The manning function is provided by the personnel sections throughout the division. The DISCOM organization shown in Figure 2-2 identifies DISCOM units in support of maneuver brigades and to the division.

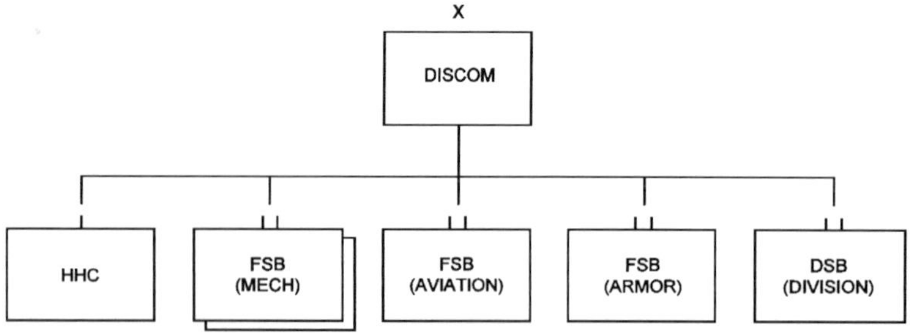

Figure 2-2. Division support command organization, Force XXI.

Division support command medical assets are organic to all of the subordinate battalions except for the division aviation support battalion (DASB). These medical assets include—

- Medical operations branch, DISCOM headquarters.
- Medical materiel management branch, DISCOM headquarters.
- Division support battalion, CHS cell.
- Division support medical company.
- Forward support battalion, CHS cell.
- Forward support medical company.

a. Medical Operations Branch (Division Support Command Headquarters). The medical operations branch is assigned to the distribution management center (DMC). The DMC has four branches—plans branch, operations branch, procurement branch, and medical operations branch. The DMC provides the division support operations the overall total asset visibility and the in-transit visibility of all commodities, movements, and units within, assigned, or inbound to the division AO. The primary responsibility of the medical operations branch is to assist with synchronization of the division CHS plan. The medical operations branch works with the DSS and the DISCOM staff in assisting with the development of the division CHS plan. This branch briefs the DISCOM commander and staff on CHS initiatives, as required. The medical operations branch and the DSS plan and coordinate for the employment of division medical assets and relocation of DISCOM CHS elements. The medical operations branch coordinates the CHS plan with synchronization of the division CHS plan. This branch collects medical information of intelligence value from reporting medical assets and forwards it to the appropriate division and DISCOM staff elements. The medical operations branch coordinates the placement of DS corps medical assets with supported DISCOM units, either in the DSA or BSA. The branch is responsible for—

- Briefing the DISCOM commander on the CHS planning and operations required.
- Providing current information that will assist the DSS with development of staff estimates and the division CHS plan.
- Coordinating the attachment of corps medical units/elements with DISCOM units.
- Providing information to the DSS on the DISCOM commander's intent for logistics and CHS operations.
- Coordinating the division CHS plan with all DISCOM staff elements.
- Coordinating operations information with the DSS and making recommendations to ensure synchronization of CHS activities in support of the division.

- Ensuring CHS information from the DSS is staffed to all DISCOM elements in a timely manner.

- Coordinating with the BSS, as required, for synchronization of division CHS.

b. Medical Materiel Management Branch. The MMMB is assigned to the general supply office (GSO). The GSO coordinates and supervises the supply management for water and Classes I, II, III(B), and III(P) supplies and recommends priorities for the allocation and other controls of supplies. The MMMB provides advice on the receipt, storage, and distribution of supplies within its area of responsibility. This office consists of a Class I branch, Classes II/III(P)/IV branch, Class III and water supply branch, Class V branch, Class VIII branch (MMMB), and Class IX branch. The MMMB manages the Class VIII supply system in the division. The branch coordinates and recommends the prioritization of medical supplies and blood products. It also coordinates for the disposition of captured enemy medical materiel. Under the technical control of the HSMO of the DSS, the MMMB monitors and coordinates Class VIII resupply for division medical units/elements. Using the CHL functional module of the TMIP/MC4 system, the Theater Army Medical Management Information System (TAMMIS), Joint Total Asset Visibility, Transportation Coordinators' Automated Information for Movement System II, and/or other automated logistics management systems, the MMMB manages all Class VIII requisitions submitted from the division to the supporting MEDLOG company. The MMMB maintains a record of the requisition until it is filled. The MMMB coordinates shortfalls in throughput distribution with the DSS and division support operation section. The MMMB may update priorities with the MEDLOG company to correct deficiencies in the delivery system. The MMMB provides Class VIII situational understanding to the DISCOM staff and the DSS according to the TSOP. For definitive information on division Class VIII resupply operations, see FMs 4-02.1, 4-02.21, 8-10-9, and 63-2-2. The MMMB, in coordination with the CHL cell of the DSS, manages the distribution of blood and blood products for division medical units. It also coordinates through the DSS with the G5 for disposition of captured enemy medical material.

c. Division Support Battalion Health Services Support Officer. The HSSO is assigned to the command section and is the medical plans and operations officer. The HSSO coordinates internal medical support. He coordinates the schedules, locations, and capabilities of medical support with the DSMC. He prepares and provides an area medical plan to the subordinate units. The HSSO is also responsible for coordinating the placement of supporting corps medical elements attached to the DSB within the battalion's assigned area of the DSA. For additional information on the DSB HSSO, see FM 63-23-2.

d. Division Support Medical Company. The DSMC has the overall mission of providing Echelons I and II CHS to units located in the DSA and division rear areas. It provides C2 for organic elements and attached medical units. The DSMC is dependent on appropriate elements of the corps and division for patient evacuation (including air ambulance), CHS operations planning, guidance, legal, finance, and personnel and administrative services. It is also dependent on the headquarters and headquarters detachment of the DSB for food service and religious support. The DSMC is organized into a company headquarters, a treatment platoon, an ambulance platoon, an optometry section, a PVNTMED section, and a MH section.

e. Forward Support Battalion Combat Health Support Cell. The CHS cell is assigned to the support operations section. It is staffed with a medical planner/HSSO who is the FSB commander's special staff officer for CHS and a member of the FSB battle staff and a medical operations NCO who is the primary assistant to the HSSO. This cell is responsible for—

- Providing the CHS input for logistics preparation of the battlefield for the FSB.

- Providing the CHS estimates and medical threat input for inclusion in the FSB commander's estimate.

- Coordinating and synchronizing FSB medical operations for the supported brigade.

- Coordinating the delivery of Class VIII supplies via logistics packages (LOGPAC).

- Overseeing all FSB CHS planning activities to ensure such planning is synchronized laterally and vertically.

- Developing the CHS portion of the FSB's OPLAN in coordination with the FSB staff, the FSMC commander, and the DISCOM medical operations branch.

- Coordinating the placement of supporting corps medical elements attached to the FSB within the BSA. Identifying CHS support requirements for the BSA including space requirements for the FST (tents, equipment, vehicles, and trailers), the forward support MEDEVAC team (FSMT) (includes fuel truck and forward area refueling equipment [FARE], aircraft parking, tents for billeting), and the supporting corps ground ambulance teams and ambulances.

- Coordinating communications access for supporting corps elements as required.

- Coordinating through the BSS with the brigade S3 (Air) for current Army airspace command and control (A2C2) information that is provided to the FSMT crews. Also provides a copy of the brigade OPORD/OPLAN A2C2 annex that provides the air corridors for medical evacuation.

- Coordinating CHS taskings from the DISCOM medical operations branch with the FSB staff and the FSMC commander. Tasking may include area medical/dental, PVNTMED, CSC, and CHS reinforcement, or reconstitution support.

- Coordinating for the training and use of nonmedical personnel for patient decontamination in the event of an NBC or weapons of mass destruction attack. (See FM 8-10-7.)

- Coordinating and synchronizing CHS requirements with the BSS and the DISCOM medical operations branch.

- Monitoring the status of the FSB and brigade medical elements via the medical situational reporting on Force XXI Battle Command Brigade and Below (FBCB2).

- Monitoring the status of division medical units/elements via medical reporting on CSSCS.

- Advising the FSB commander on CHS operations in the BSA and brigade rear.

- Maintaining situational understanding of lateral and supporting medical units.

- Submitting and forwarding status reports in accordance with DISCOM and brigade TSOP.

f. Forward Support Medical Company. The FSMC has the overall mission of providing Echelon I and Echelon II CHS on a DS basis for the supported maneuver brigade. It provides C2 for organic elements and attached medical units. The FSMC is dependent on appropriate elements of the corps, division, brigade, and FSB for patient evacuation (including air ambulance), CHS operations planning and guidance, and for legal, finance, and personnel and administrative services. It is also dependent on the headquarters and distribution company of the FSB for food service and religious support and the base support company for maintenance. The FSMC is organized into a company headquarters, a treatment platoon, an ambulance platoon, a PVNTMED section, and a MH section (see Figure 2-3). For more detailed information on the operations and functions of the medical company, see FM 8-10-1. The FSMC performs these functions:

- Treatment of patients with DNBI, battle fatigue (BF), and trauma injuries. It provides routine sick call, triage of mass casualties, ATM, surgical resuscitation/stabilization (when the FST from the corps is deployed/collocated with the FSMC), and preparation of patients incapable of returning to duty for further evacuation.

- Ground ambulance evacuation for patients from BAS and designated casualty collection points (CCP).

- Operational dental care (emergency and essential dental care).

- Class VIII resupply and medical equipment maintenance for supported units.

- Medical laboratory and radiology services commensurate with Echelon II/division-level treatment.

- Outpatient consultation services for patients referred from unit-level MTF.

- Patient holding for up to 40 patients able to RTD within 72 hours.

- Limited reinforcement and augmentation to supported maneuver battalion medical platoons.

- Coordination with the UMT for required religious support.

- Preventive medicine consultation and support.

- Combat stress control, to include management of BF and stress-related casualties.

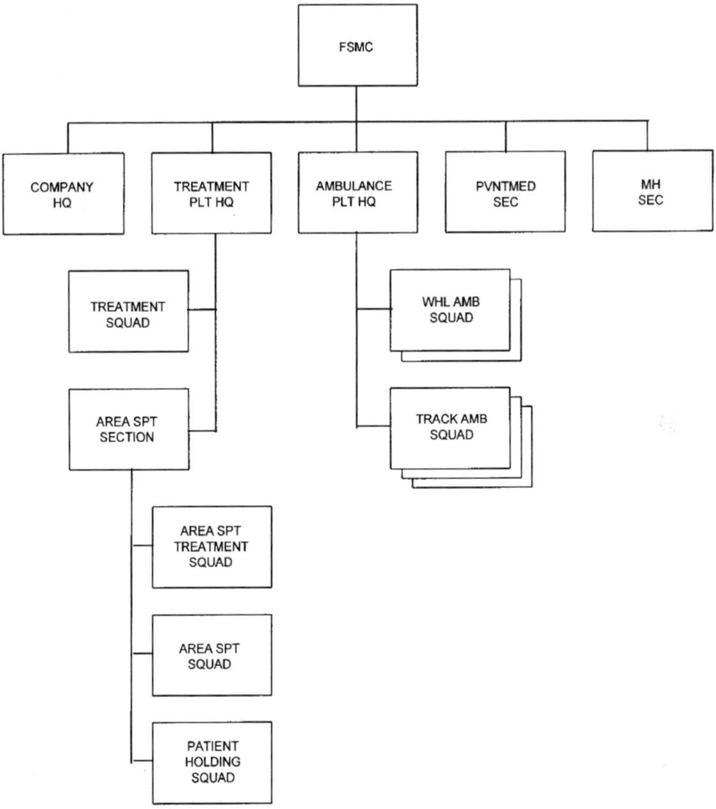

Figure 2-3. Forward support medical company.

Section III. BRIGADE HEADQUARTERS MEDICAL ASSETS

2-10. Brigade Surgeon, F-Edition TOE (Force XXI Digitized Division)

The brigade surgeon is an MC officer (Major, AOC 62B00). He is a special staff officer who plans and coordinates brigade CHS activities with the brigade staff. The brigade surgeon is assigned to the HHC of the maneuver brigade. The surgeon is responsible for the technical control of all medical activities in the command. The brigade surgeon oversees and coordinates CHS activities through the BSS and the brigade S3. The brigade surgeon keeps the brigade commander informed on the status of CHS for brigade operations and the health of the command. He provides input and obtains information to facilitate medical planning. His specific duties in this area include—

- Ensuring implementation of the CHS section of the brigade TSOP.
- Determining the allocation of medical resources within the brigade.
- Supervising technical training of medical personnel and the CLS program within the brigade.
- Determining procedures, techniques, and limitations in the conduct of routine medical care, emergency medical treatment (EMT), and ATM.
- Monitoring aeromedical and ground ambulance evacuation.
- Monitoring the implementation of automated medical systems.
- Informing the division surgeon on the brigade's CHS situation.
- Monitoring the health of the command and advising the commander on measures to counter disease and injury threats.
- Exercising technical supervision of subordinate battalion surgeons and PAs.
- Providing consultation and mentoring for subordinate battalion surgeons, physicians, and PAs.
- Providing the medical estimate and medical threat for inclusion in the commander's estimate.
- Monitoring the command PVNTMED program (includes health assessment and medical surveillance); see AR 40-5 and FM 4-02.17.
- Ensuring field health records are maintained by primary care providers according to AR 40-66.

2-11. Brigade Surgeon's Section, F-Edition TOE (Force XXI/Digitized Division)

The BSS is assigned to the HHC of the brigade and operates out of the brigade TOC. The section, in coordination with the HSSO of the FSB support operations section and the FSMC commander, is responsible for the development of the medical portion of the brigade OPLAN/OPORD and takes part in the brigade planning process. The BSS staff is responsible to the brigade commander for staff supervision of CHS within the brigade. The BSS is also responsible for coordinating GS and DS relationships of organic medical units and medical units/elements whether under OPCON or attached to the brigade. The brigade commander is updated as required on the status of CHS in the brigade. Figure 2-4 shows the typical organization and staffing of the BSS. It consists of a medical plans and operations cell and a patient disposition and reports cell. The staff of the BSS assists the brigade surgeon in planning and conducting brigade CHS operations.

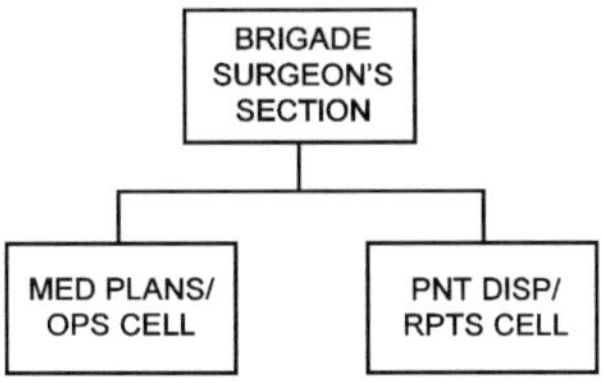

Figure 2-4. Brigade surgeon's section.

2-12. Brigade Surgeons, L-Edition TOE (Army of Excellence/AirLand Battle)

The FSMC commander positions are documented 05A, AMEDD immaterial, meaning any qualified AMEDD officer can assume command. When the FSMC commander is not a physician, medical decisions and technical supervision of physicians is performed by the senior physician/treatment platoon leader. When a brigade surgeon is not assigned to the brigade headquarters, the treatment platoon leader, who is always a physician, will perform the brigade surgeon's duties. His duties and responsibilities as the brigade surgeon require that he work closely with the FSMC commander and include, but are not limited to—

- Ensuring the implementation of the CHS section of the division TSOP.
- Determining the allocation of CHS resources within the brigade.
- Supervising the technical training of medical personnel and the CLS program within the brigade.
- Developing and monitoring the MEDEVAC plan (ground and air) which supports the brigade's maneuver plan.
- Writing the CHS portion of brigade TSOP, OPLAN, and OPORD.
- Monitoring requests for aeromedical evacuation from supported units.
- Monitoring the health of the command and advising the commander on measures to counter the medical threat.
- Monitoring and advising units on their mild to moderate BF cases and determining the capability to restore BF casualties within the brigade's AO.

- Informing the division surgeon and the DMOC of the brigade's CHS situation.

- Supervising corps medical elements within the brigade's AO when directed.

- Exercising technical control over subordinate battalion surgeons.

- Assuming technical supervision of PAs organic to subordinate units in the absence of their assigned physicians.

- Advising PAs assigned to artillery and engineer battalions, as required.

2-13. Armored Cavalry Regiment and Separate Brigade Surgeons (Army of Excellence)

The duties of the regimental surgeon and the separate brigade surgeon are the same as those identified in paragraph 2-10.

2-14. Armored Cavalry Regiment Medical Troop

The mission of the ACR medical troop is to provide Echelons I and II medical care within the ACR. The capabilities of this unit are to—

- Provide C2 of attached medical elements (including CHS planning; policies and procedures; support operations; and MEDEVAC coordination for movement of patients within and out of the regiment AO).

- Advise the regiment commander and support squadron commander on the health of the command and other CHS activities affecting the regiment.

- Develop, prepare, and coordinate the CHS portion of OPLAN and OPORD.

- Allocate medical resources (personnel and equipment) to all assigned and attached units of the regiment.

- Perform triage, initial resuscitation and stabilization, and preparation for further evacuation of patients generated in the regiment rear area.

- Provide ground evacuation for patients from Echelon I MTFs.

- Employ treatment squads to perform reinforcement/augmentation to maneuver squadrons' medical platoons. (These squads/teams are routinely placed OPCON to supported maneuver squadrons.

They are normally attached to the squadron medical platoon under technical control of the squadron surgeon.)

- Provide CHL and medical equipment maintenance repair parts and support to the regiment on an area support basis. (The regiment medical supply section maintains a 5-day stock of emergency *push* packages and individual medical items. Emergency supply requests are sent to the supporting MEDLOG battalion or the nearest medical unit.)

- Provide dental support (including treatment of maxillofacial injuries; operational dental care that includes emergency and essential dental treatment).

- Provide laboratory service commensurate with the regiment's Echelon II facility.

- Perform patient holding for up to 40 patients awaiting evacuation or RTD within 72 hours.

- Provide outpatient consultation services for patients referred from Echelon I MTFs.

Section IV. MEDICAL PLATOONS (ARMY OF EXCELLENCE/AIRLAND BATTLE)

2-15. Assignment

A medical platoon is organic to each combat battalion HHC. Under the Army of Excellence TOE, the platoon is organized with a headquarters section, a treatment squad (two treatment teams), an ambulance squad, and a combat medic section. The medical platoon is organized as shown in Figures 2-5 and 2-6.

NOTE

Mechanized infantry and armor units have four ambulance squads consisting of two ambulance teams each assigned to their medical platoon ambulance squads. Airborne and air assault medical platoons have 12 trauma specialists assigned to their combat medic section, while light infantry has 9 trauma specialists assigned to their combat medic sections. The armor medical platoons have 3 trauma specialists assigned to their combat medic section and the mechanized infantry medical platoons have 3 health care SGTs (91W30) and 9 trauma specialists (91W10) assigned to their combat medic section.

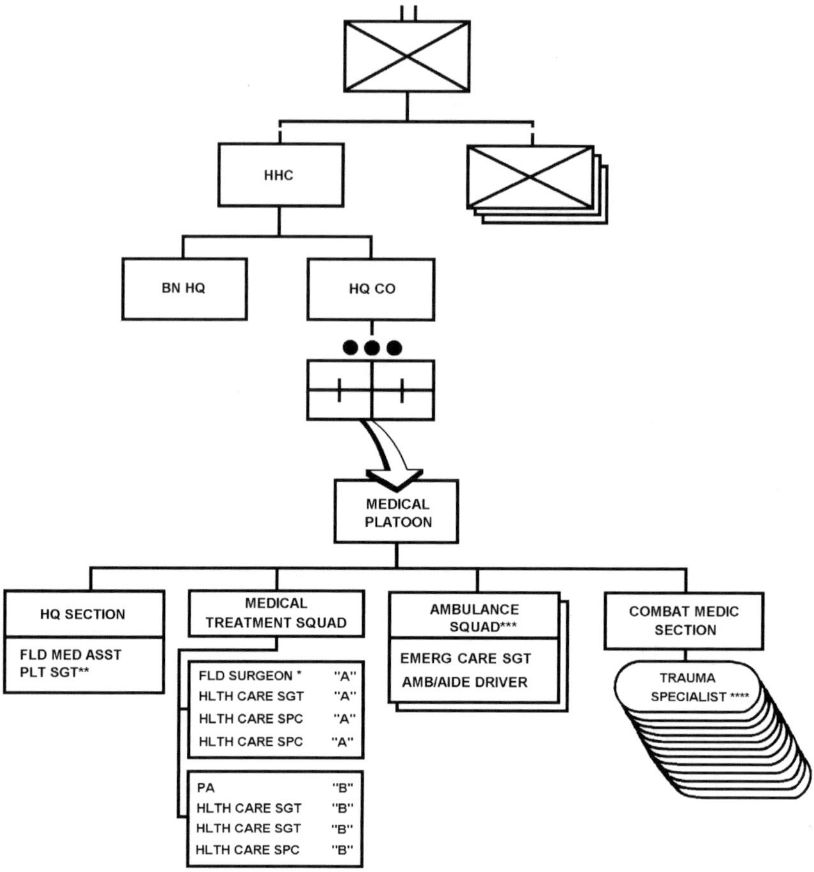

* ALSO BATTALION SURGEON
** ALSO SERVES AS AMBULANCE SECTION SGT
*** 4 AMBULANCE TEAMS
**** AIRBORNE BATTALIONS HAVE 12 TRAUMA SPECIALISTS ASSIGNED TO THE COMBAT MEDIC SECTION AND AIR ASSAULT BATTALIONS HAVE 10 TRAUMA SPECIALISTS ASSIGNED TO THE COMBAT MEDIC SECTION
"A" ALFA TREATMENT TEAM
"B" BRAVO TREATMENT TEAM

Figure 2-5. Medical platoon, headquarters and headquarters company light infantry battalion.

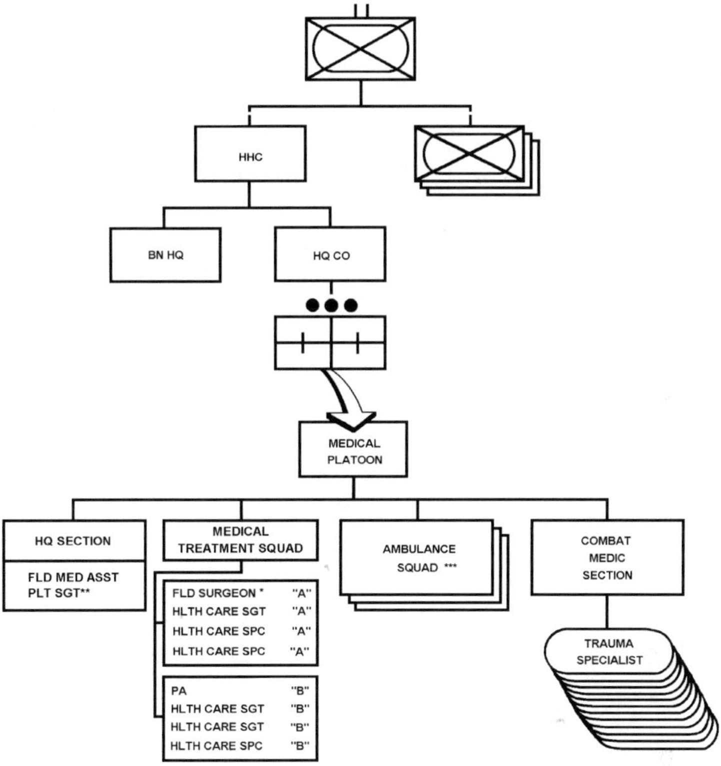

* ALSO BATTALION SURGEON
** ALSO SERVES AS AMBULANCE SECTION SGT
*** MECHANIZED INFANTRY BATTALIONS HAVE 3 AMBULANCE SQUADS AND ARMOR BATTALIONS HAVE 4 AMBULANCE SQUADS
"A" ALFA TREATMENT TEAM
"B" BRAVO TREATMENT TEAM

Figure 2-6. Medical platoon, mechanized infantry battalion.

2-16. Battalion Surgeon/Medical Platoon Leader, L-Edition TOE

The battalion surgeon/medical platoon leader (MC, AOC 62B00) is the medical advisor to the battalion commander and his staff. He is the supervising physician (operational medicine officer) of the medical platoon treatment squad. This officer is responsible for all medical treatment provided by the platoon. His responsibilities include—

- Planning and directing Echelon I CHS for the battalion.
- Advising the battalion commander and his staff on the status of the health of the command.
- Supervising the administration, discipline, maintenance of equipment, supply functions, organizational training, and employment of assigned or attached personnel.
- Examining, diagnosing, treating, and prescribing courses of treatment for patients, to include ATM.
- Coordinating the establishment and training of nonmedical personnel for patient decontamination teams.
- Training CLS.
- Supervising the battalion MH/CSC program, to include training troop leaders in the preventive aspect of stress on soldiers.
- Supporting humanitarian assistance programs when directed.
- Overseeing the common task training, continuing medical education, and clinical training of subordinate medical personnel.
- Monitoring the command PVNTMED program, to include health assessment and medical surveillance; see AR 40-5 and FM 4-02.17.
- Ensuring field health records are maintained by primary care providers according to AR 40-66.

2-17. Platoon Headquarters, L-Edition TOE

a. The headquarters section, under the direction of the battalion surgeon/medical platoon leader, provides for the C3 and resupply for the platoon. The platoon headquarters is manned by the field medical assistant and the platoon SGT. It is normally collocated with the treatment squad to form the BAS. The CP includes the plans and operations functions performed by the field medical assistant. The platoon has access to the battalion wire communication network for communications with all major elements of the battalion and with supporting units. Wireless communications for this section consists of a tactical FM radio mounted in the platoon headquarters vehicle. The medical platoon employs an FM radio network for CHS operations (Figure 2-7). The headquarters section serves as the net control station (NCS) for the platoon.

b. The field medical assistant, an MS Corps officer, is the operations/readiness officer for the platoon. He is the principal assistant to the battalion surgeon/medical platoon leader for operations, administration, and logistics. The field medical assistant coordinates CHS operations with the battalion S3 and S4 and coordinates patient evacuation with the FSMC.

c. The platoon SGT assists the platoon leader and supervises the operations of the platoon. He also serves as the ambulance section SGT. This NCO prepares reports; requests general supplies as well as medical supplies; advises on supply economy procedures; and maintains authorized stockage levels (ASL) of expendable supplies. He supervises the activities and functions of the ambulance section, to include operator maintenance of ambulances and equipment; OPSEC; and EMT.

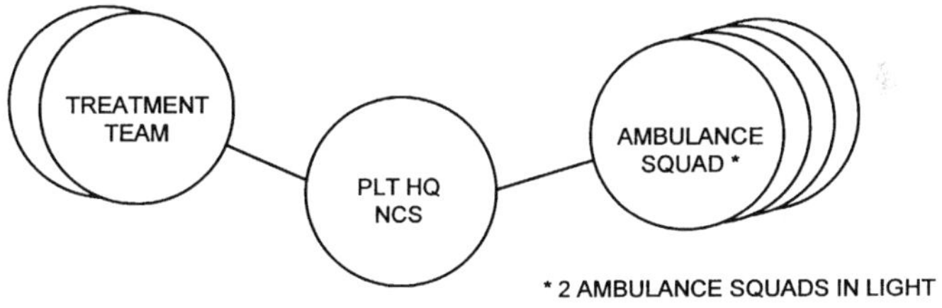

Figure 2-7. Medical platoon operations net.

d. The PA performs general technical health care and administrative duties. The PA is ATM-qualified and works under the clinical supervision of the medical officer. This officer serves as the medical platoon leader in the absence of an assigned physician. He performs the following duties:

- Establishes and operates a BAS or BAS minus (one treatment team).
- Treats, within his ability, sick or injured patients. He refers those patients requiring treatment beyond his capability to the supervising physician.
- Provides initial resuscitation to wounded personnel, as required.
- Conducts training for battalion personnel in first aid procedures (self-aid/buddy aid), CLS, field sanitation, evacuation of the sick, injured, and wounded, and the medical aspects of injury prevention. For additional information on the CLS Program, see Appendix C.
- Assists in the conduct of the battalion preventive psychiatry program, to include training troop leaders in the preventive aspects of stress on soldiers.

- Trains medical personnel in emergency medical procedures and, in the absence of a physician, ensures common task training, continuing medical education, and clinical training of subordinate medical personnel.

2-18. Treatment Squad, L-Edition TOE

The treatment squad is the basic medical treatment element of the BAS. It provides routine sick call services, emergency medical care, triage, and ATM. This squad is staffed with an operational medicine officer (primary care physician/battalion surgeon), a PA, two health care SGTs, and four health care specialists. The squad's physician and PA are trained in ATM procedures.

2-19. Battalion Aid Station/Treatment Squad Operations, L-Edition TOE

Battalion aid station is the generic term used in designating the unit-level/Echelon I MTF.

a. The treatment squad can split into two treatment teams and operate as two separate aid stations (BAS minus), normally not to exceed 24 hours. In continuous operations, when operating for longer periods, personnel efficiency and unit capability will tend to deteriorate. Each team employs treatment vehicle(s) with two medical equipment sets (MES)—one trauma field MES, and one sick call field MES.

b. For communications, each treatment team uses an FM tactical radio and is deployed in the medical platoon's operations net. However, under certain tactical conditions, the battalion S4 may require BAS elements to use the S4 net.

c. The BAS is under the tactical control of the battalion S4 and is normally deployed in the vicinity of combat trains (see Figures 2-8 and 2-9 for suggested layout of a BAS). To reduce ambulance turnaround time in providing ATM to patients within 30 minutes of wounding, the BAS may split and place its treatment teams as close to maneuvering companies as tactically feasible. The battalion S4 closely coordinates locations for forward positioning CSS elements (including medical treatment elements) with the battalion S3. This is to ensure that the location of these elements is known by commanders of maneuvering forces. Coordination ensures that CSS elements are not placed in the way of friendly maneuvering forces, in line of fires, or in areas subject to be overrun by rapidly advancing enemy forces. Treatment teams situated close to (within 1,000 meters of) maneuvering companies in contact must be prepared to withdraw to preplanned, alternate positions on short notice.

d. When maneuvering companies anticipate large numbers of casualties, augmentation of the medical platoon with one or more treatment teams from the FSMC should be made. Augmenting treatment teams are under the tactical control of the battalion S4; but are under the OPCON of the battalion surgeon/ medical platoon leader. A suggested scheme of employment is to place a team in close support of each maneuvering company while locating one treatment team in the combat trains. Medical treatment facilities should not be placed near targets of opportunity such as ammunition, petroleum, oils and lubricants (POL), distribution points, or other targets that may be considered lucrative by the opposing force. Considerations for the location of the BAS should include—

- Tactical situation/commander's plan.
- Expected areas of high casualty density.
- Security.
- Protection afforded by defilade.
- Convergence of lines of drift.
- Evacuation time and distance.
- Accessible evacuation routes.

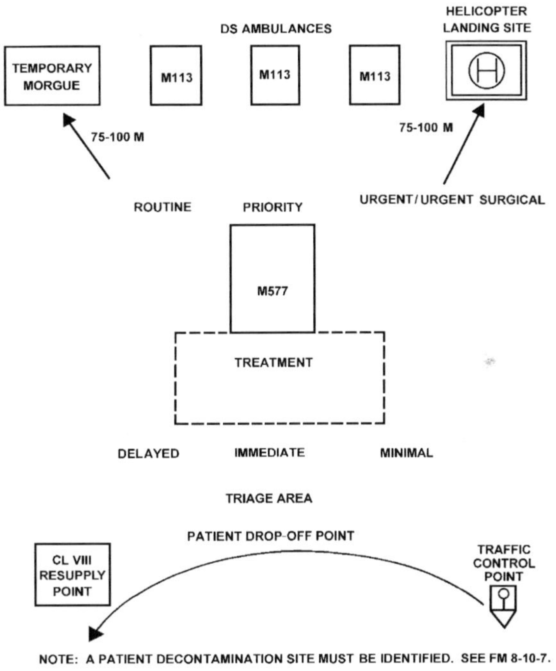

Figure 2-8. Layout of a battalion aid station (heavy).

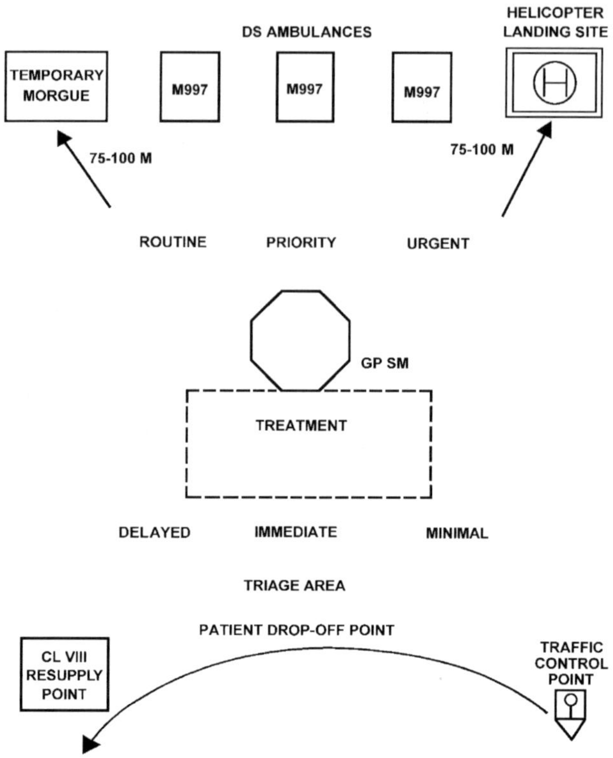

Figure 2-9. Layout of a battalion aid station (light).

- Avoidance of likely target areas such as bridges, fording locations, road junctions, and firing positions.

- Solid ground with good drainage.

- Near an open area suitable for helicopter landing.

- Available communication means.

- Additional space near this site for establishing a patient decontamination site if required.

e. At the BAS, patients requiring further evacuation to the rear are stabilized for movement. Constant efforts are made to prevent unnecessary evacuation; patients with minor wounds or illnesses are treated and RTD as soon as possible. Other functions of the BAS include—

- Receiving and recording patients.
- Notifying the S1 of all patients processed through the BAS, giving identification and disposition of patients.
- Preparing Field Medical Cards (FMC) (Department of Defense [DD] Form 1380), as required. See FM 8-10-6 for information on completion and disposition of the form.
- Verifying information contained on the FMC of each patient evacuated to the BAS.
- Requesting and monitoring MEDEVAC of patients.
- Monitoring personnel, when necessary, for NBC contamination prior to medical treatment.
- Supervising patient decontamination and treating NBC patients (refer to FMs 8-10-7, 4-02.283, 8-284, and 8-285).

NOTE

Patient decontamination is performed by a pretrained team. This team is composed of eight nonmedical personnel from supported units working under the supervision of medical personnel. Patient decontamination teams perform best when they train and exercise their skills with the supporting BAS.

f. Medical evacuation from the BAS is performed by ground ambulances from the FSMC and by corps air ambulance teams.

g. Patient holding and food service is not available at the BAS. Therefore, only procedures necessary to preserve life or limb, or enable a patient to be moved safely, are performed at the BAS.

h. Ammunition and individual weapons belonging to patients evacuated from the BAS are disposed of as directed by command standing operating procedures (SOP)/policy. All excess equipment collected at the BAS is disposed of by the battalion S4 or as directed by command SOP.

NOTE

Patients will always retain their protective mask when evacuated to the next echelon of care, as long as they are in the combat zone. Based on the threat, they may retain the protective mask until evacuated out of the theater.

i. Patients requiring dental treatment are provided relief for dental pain, if required, then evacuated to the supporting medical company where operational dental care (emergency and essential dental treatment) is provided.

j. Patients requiring optometry services initially report to the BAS. For those patients requiring only routine replacement of spectacles, necessary information is obtained from the individual and forwarded to the division optometry section. The required spectacles are fabricated and forwarded to the BAS for issue to the patient. For optometry services other than routine repair or replacement of spectacles, patients are transported to the optometry section, MSMC, located in the DSA.

2-20. Combat Medic Section, L-Edition TOE

To foster good interpersonal relations and morale of combat troops, every effort should be made to attach the same trauma specialists to the same unit they habitually support each time the unit deploys. However, during lulls in combat operations, they should return to the medical platoon for consultation and proficiency training. Functions of trauma specialists are as follows:

- Performs triage and EMT for the sick and wounded.
- Arranges MEDEVAC for litter patients and directs ambulatory patients to CCP or to the BAS.
- Initiates a FMC for the sick and wounded and, as time permits, prepares a FMC on deceased personnel.
- Screens, evaluates, and treats, within his capabilities, those patients suffering minor illnesses and injuries.
- Keeps the company commander and the battalion surgeon/medical platoon leader informed on matters pertaining to the health and welfare of the troops.
- Manages Class VIII resupply for the unit's CLS.
- Maintains sufficient quantities of medical supplies to support the tactical situation.
- Serves as a member of the unit field sanitation team. In this capacity, he advises the commander and supervises unit personnel on matters of personal hygiene and field sanitation (FM 21-10-1).

2-21. Ambulance Squads, L-Edition TOE

a. Medical platoon ambulance squads provide evacuation within the battalion. Ambulance teams provide medical evacuation and en route care from the soldier's point of injury or a CCP to the BAS. In mass casualty situations, nonmedical vehicles may be used to assist in casualty evacuation (CASEVAC) as directed by the commander. Plans for the use of nonmedical vehicles to perform CASEVAC should be included in the battalion's TSOP.

b. Under the modular medical system, the ambulance squad consists of two ambulance teams.

(1) The emergency care SGT—

- Performs triage and EMT procedures in the care and management of trauma patients.
- Assists in the care and management of BF patients.
- Prepares patient for movement.
- Provides patient care en route.
- Maintains contact with supported units.
- Collects casualties.
- Performs NBC detection procedures.

(2) The ambulance/aide driver is trained in EMT procedures. He operates and maintains the ambulance and all onboard equipment. He assists the aide/evacuation NCO in the care and handling of patients.

c. Specific duties of the ambulance team are to—

- Maintain contact with supported elements.
- Find and collect the wounded.
- Administer EMT as required.
- Initiate or complete the FMC.
- Evacuate litter patients to the BAS.
- Direct or guide ambulatory patients to the BAS.
- Perform triage when necessary.

- Provide Class VIII resupply to trauma specialists.
- Serve as messengers within medical channels.
- Maintain operational readiness of assigned vehicle.

d. The number of ambulance squads in a section varies and is based on the type of parent organization. The infantry, airborne, and air assault maneuver battalions' ambulance sections have two ambulance squads; each is equipped with HMMWV ambulances. The mechanized infantry and the armored battalions ambulance sections have four ambulance squads equipped with M-113 tracked ambulances.

2-22. Employment and Functions of the Ambulance Team, L-Edition TOE

a. The ambulance team is a mobile trauma specialist team. Its function is to collect, treat, and evacuate the sick and wounded to the nearest treatment station or ambulance exchange point (AXP). For communications, the ambulance team employs an FM tactical radio mounted on its assigned ambulance. The team uses the medical platoon's internal operations net; however, in certain circumstances it may operate in the battalion administration/logistics net or as established by the battalion signal operating instructions (SOI).

b. In the heavy maneuver battalions, the track ambulance team routinely deploys with the maneuver company trains; however, it operates as far forward as the tactical situation permits and evacuates patients from the point of injury, if possible. In the light maneuver battalion, the wheeled ambulance team is either dispatched from the BAS, pre-positioned as close to the supported units as the tactical situation permits, or positioned with the maneuver company trains. Ambulance teams operating in a company's AO are normally under the tactical control of the company XO or 1SG, but remain under the technical and OPCON of the medical platoon. An ambulance team from the BAS will habitually support the same company. To become familiar with the specific terrain and battlefield situation, the team maintains contact with the company during most combat operations.

c. During static situations where the company is not in enemy contact or is in reserve, the team returns to the BAS to serve as backup support for other elements in contact. However, during movement to contact, the ambulance team immediately deploys to its regularly supported company. During combat operations, the team may dismount (leaving the ambulance in the company trains area), find, treat, and move patients to safety, and later evacuate them to the BAS. When moving patients to the ambulance location, CCP, or company aid post, the team is normally assisted by nonmedical personnel.

2-23. Medical Evacuation, L-Edition TOE

a. Optimum patient care and treatment is dependent upon an evacuation system that provides a continuous movement of patients. Medical evacuation is the process of moving patients from the point of injury or illness to an MTF, while providing en route medical care, or between MTFs. Each stop in the process is to provide medical treatment to enhance the patient's early RTD or to stabilize him for further evacuation. The responsibility for patient evacuation rests with the echelon of CHS to which the patient is to be evacuated (see Figure 2-10). Ambulances go forward, pick up patients, and move them to the supporting MTFs.

(1) Ambulance teams of the medical platoon evacuate patients from the company aid post or CCP to the BAS.

(2) Ambulance squads of the FSMC evacuate patients from the BAS to the division clearing station.

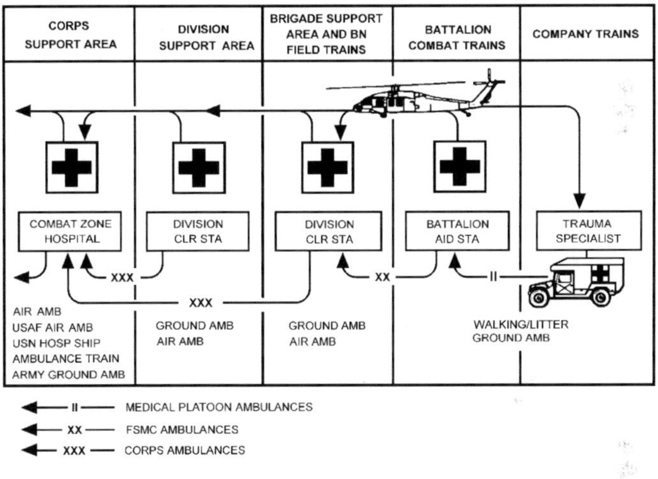

Figure 2-10. Patient evacuation flow.

b. An ambulance shuttle system may be set up between the FSMC division clearing station and the BAS. An AXP is established (Figure 2-11) so that ambulances are moving forward as others move rearward; thus enabling a continuous rearward evacuation flow, while decreasing ambulance turnaround time. Patients are evacuated no further to the rear than their conditions require.

c. Aeromedical evacuation in the combat zone should be used to the maximum extent possible for critically ill or wounded patients. See FMs 8-10-6 and 8-10-26 for additional information on aeromedical evacuation. Refer to FM 8-10-6 for MEDEVAC request procedures. Normally, ground ambulances are used to evacuate the minimally ill or wounded and those patients who cannot be evacuated by air. The specific mode of evacuation is determined by the patient's condition, aircraft/vehicle availability, the tactical situation, and weather conditions (METT-TC factors). When both air and ground ambulances are used, specific factors are considered in determining which patients are to be evacuated by air and which are to be evacuated by ground ambulances (see FM 8-10-6). Normally, the physician or PA treating the patient (or the senior trauma specialist in their absence) makes this determination; it is based on the medical condition of the patient. However, the goal is to get the trauma patient to the initial treatment/ATM element within 30 minutes of wounding.

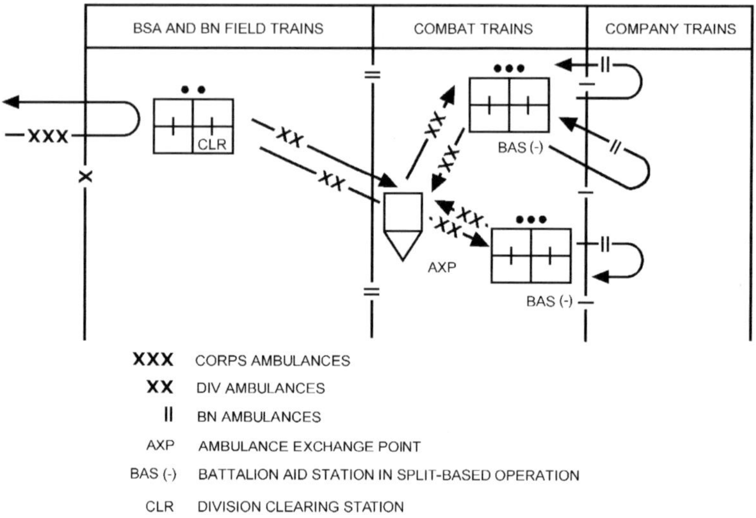

Figure 2-11. Ground ambulance shuttle system.

2-24. Evacuation and Disposition of Remains

a. The transportation and disposition of remains is a Quartermaster function. Air and ground ambulance personnel do not clear the battlefield of remains nor do they carry remains in their dedicated medical vehicles or aircraft. Medical units do not accept remains or provide temporary morgues in which to hold remains for other units. Other units are responsible for evacuation of remains to mortuary affairs collection points.

b. The only remains that medical units/elements handle are those of its own unit members or of patients who are dead on arrival (DOA) or who died of wounds (DOW) while in their care. Whenever a medical unit/element establishes a temporary morgue, it should be out of sight of the triage and treatment areas. The temporary morgue/holding area can be established behind a natural barrier, such as a stand of trees or it can be shielded from the view of others by using either tents or tarpaulins.

2-25. Class VIII Resupply, L-Edition TOE

a. The medical platoon maintains a 2-day (48-hour) stockage of Class VIII supplies within its MES. The following MES are authorized for the medical platoon treatment section and they include—

- Chemical Agent Patient Decontamination, National Stock Number (NSN) 6545-01-176-4612 (1).
- Chemical Agent Patient Treatment, NSN 6545-01-141-9469 (2).
- Sick Call Field, NSN 6545-01-228-1886 *(2).
- Trauma Field, NSN 6545-01-228-1667 *(2).

* Indicates the numbers of MES authorized for each treatment team.

Normal medical resupply of the platoon is performed by the DMSO through LOGPACs, backhaul, or in coordination with the movement control office(r) (MCO). Medical resupply may also be by preconfigured Class VIII packages (*push* packages) throughput from the MEDLOG battalion located in the corps support area (Figure 2-12).

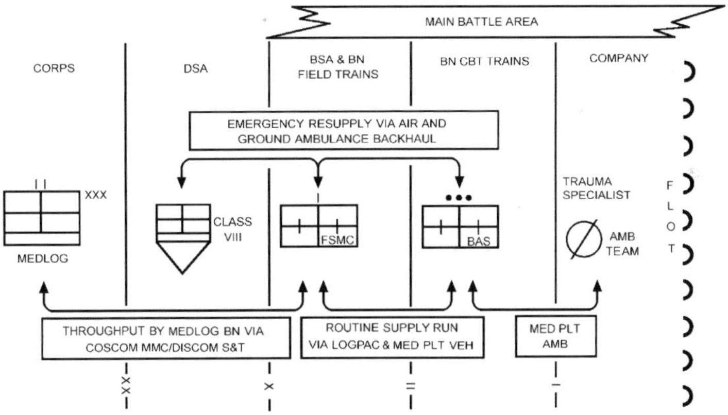

Figure 2-12. Flow of Class VIII supplies.

b. In a tactical environment, the emergency medical resupply (ambulance backhaul) system is used. In this environment, medical supplies are obtained informally and as rapidly as possible, using any available medical transportation assets. The medical platoon submits supply requests to the supporting FSMC, who in turn fills requests and ships supplies forward. Request for items not available at the FSMC are forwarded to the DMSO; the request is filled from division stocks and shipped to the requestor by the most expedient means available. Air ambulances from corps and ground ambulances from the DISCOM transport medical supplies directly to BAS. Class VIII resupply of trauma specialists are performed by ambulances of the medical platoon. The trauma specialist can also be resupplied from the ambulance crew from supplies onboard the ambulance.

Section V. MEDICAL PLATOONS, FORCE XXI/DIGITIZED DIVISION

2-26. Medical Platoon Assignment, F-Edition TOE

The medical platoon is organic to all maneuver battalions. In the armored battalions and mechanized infantry battalions, the platoon is organized with a headquarters section, a treatment section, ambulance squads, and a combat medic section.

NOTE

1. One 91W10 per armor company and one 91W30 and three 91W10s per mechanized infantry company.

2. One ambulance team per maneuver company supported.

The medical platoon receives Echelon II CHS from the supporting FSMC.

2-27. Platoon Headquarters, F-Edition TOE

The headquarters section, under the direction of the platoon leader, provides the C3 and logistics for the platoon. The platoon headquarters is manned by the field medical assistant and the platoon SGT. It is normally collocated with a treatment team/squad to form the BAS. The CP includes the plans and operations functions performed by the field medical assistant. The platoon has access to the HHC and the maneuver battalion wire communication network for communications with all major elements of supported and supporting units. Wireless communications for this section consists of a tactical FM radio mounted in the platoon headquarters vehicle. The medical platoon employs an FM radio network for CHS operations, to include telemedicine and teleconsultation procedures. The headquarters section serves as the NCS for the platoon. Each of the medical platoon vehicles have CSS functions for the FBCB2 system. The FBCB2 is a hardware/software suite that digitizes C2 at brigade level and below. The FBCB2 concept provides a seamless battle command capability for performance of missions throughout the operational continuum at the tactical level. The FBCB2 is the implementation of information age technology to provide increased battlefield operational capabilities. The system, positioned on the ambulance and treatment vehicles, will perform combat, CS, and CSS functions for the planning and execution of operations. The FBCB2 represents a major paradigm shift for the CSS and CHS communities. For the first time, the CSS organizations are digitally linked to the platforms and organizations that they support and the CHS elements are digitally linked to brigade and FSB medical elements. The FBCB2 provides a common operations picture enabling CHS and CSS providers to maintain the operational tempo set by maneuver commanders and to have near-time situational understanding of what is taking place on the battlefield.

2-28. Battalion Surgeon/Medical Platoon Leader, F-Edition TOE

The battalion surgeon/medical platoon leader (MC, AOC 62B) is a working physician on Treatment Team "Alpha." He is the medical advisor to the supported battalion commander and his staff. He is also the supervising physician (field surgeon) of the medical platoon's treatment teams. This officer is responsible for all medical treatment provided by the platoon. His responsibilities include—

- Planning and directing CHS for the supported maneuver battalion.
- Advising the supported maneuver battalion commander and his staff on CHS operations and the medical threat.
- Supervising the administration, discipline, maintenance of equipment, supply functions, organizational training, and employment of assigned or attached personnel.
- Examining, diagnosing, and treating (or prescribing courses of treatment) for patients, to include telementoring (TMEN) and ATM.
- Training CLS.
- Supervising the battalion CSC program, to include individual and leader training on the prevention of BF and other stress-related conditions.
- Planning and conducting humanitarian assistance programs when directed.
- Coordinating the medical evacuation of patients, as required.

The field medical assistant, an MS officer, is the operations/readiness officer for the platoon. He is the principal assistant to the platoon leader for operations, administration, and logistics. The field medical assistant coordinates CHS operations with the forward support company (FSC) support operations, the supported TF S1 and S4, and MEDEVAC with the FSMC. The platoon SGT assists in supervising the operations of the platoon. He also serves as the ambulance section SGT. Physician assistants are assigned to the Bravo and Charlie treatment teams. The PA (AOC 65D) performs general technical health care and administrative duties. He is ATM-qualified and works under the clinical supervision of the medical officer. The PA performs the following duties:

- Establishes and conducts treatment team operations when deployed to other locations away from the BAS.

- Treats, within his ability, sick or injured patients. He refers those patients requiring treatment beyond his capability to the supervising physician.

- Provides ATM for wounded and injured patients.

- Provides medical treatment for DNBI patients.

- Conducts training for battalion personnel in first-aid procedures (self-aid, buddy aid, and CLS), field sanitation, evacuation of the sick and wounded, and the medical aspects of injury prevention.

- Assists in the conduct of the battalion CSC program, to include individual and leader training on the prevention of BF and other stress-related conditions.

- Trains medical personnel in EMT procedures.

2-29. Treatment Section, F-Edition TOE

The three treatment teams (Teams Alpha, Bravo, and Charlie) are the basic medical treatment elements of the BAS. They provide Echelon I medical care and treatment. This includes sick call, EMT, ATM, and triage for the management of mass casualty situations. Each treatment team is staffed with a primary care physician or a PA, one health care SGT (E-5 or E-6) and two health care specialists (E-4 or E-3). The physician, PA, and health care personnel are all trained in ATM procedures, commensurate with their positions and skill levels.

2-30. Combat Medic Section, F-Edition TOE

Trauma specialists are allocated to mechanized infantry companies on the basis of one trauma specialist per platoon and a senior health care SGT for each company. In armored units, the allocation is one health care SGT and, normally, one ambulance team per company.

a. Trauma Specialist Location. The mechanized infantry platoon trauma specialist normally locates with, or near, the element leader. When the platoon is moving on foot in the platoon column formation, he positions himself near the element leader trailing the base squad forward of the second team. This formation is the platoon's primary movement formation. When the platoon is mounted, the trauma specialist will normally ride in the same vehicle as the platoon SGT.

b. Health Care Sergeant. The company health care SGT or specialist with the armor company normally collocates with the 1SG. When the company is engaged, he remains with the 1SG and provides medical advice, as necessary. As the tactical situation allows, he will provide medical treatment and prepare patients for MEDEVAC. The ambulance team supporting the company works in coordination with the trauma specialists supporting the platoons. When a casualty occurs in a tank or an armored fighting vehicle, the ambulance team will move as close to the vehicle as possible, making full use of cover, concealment, and defilade. Assisted, if possible, by the vehicle's crew, they will extract the casualty from the vehicle and administer EMT. They move the patient to the treatment team or to a CCP to await further MEDEVAC. The company health care SGT normally remains with the company CP, but may be used anywhere in the company, assisting the ambulance teams in some situations.

2-31. Ambulance Squads, F-Edition TOE

There are three ambulance squads assigned to the medical platoon. Each squad has two ambulance teams. Armored ambulance teams have three emergency care personnel while wheeled ambulances have two emergency care personnel. The platoon ambulances provide medical evacuation within the supported maneuver battalion/TF. Ambulance teams provide medical evacuation and en route care from the soldier's point of injury to the BAS. In mass casualty situations, nonmedical vehicles may be used to assist in CASEVAC as directed by the supported commander. Plans for the use of nonmedical vehicles to perform CASEVAC should be included in the maneuver battalion's TSOP and OPORD.

2-32. Property Exchange

a. United States Army Medical Evacuation Operations. Whenever a patient is evacuated from one MTF to another or is transferred from one ambulance to another, medical items of equipment (casualty evacuation bags [cold weather-type bags], blankets, litters, and splints) remain with the patient. To prevent rapid and unnecessary depletion of supplies and equipment, the receiving Army element exchanges like property with the transferring element. This reciprocal procedure will be practiced to the fullest extent possible through all phases of evacuation from the most forward element through the most rearward hospital.

This subparagraph implements STANAG 2128 and QSTAGs 435 and 436.

b. Medical Property of Allied Nations (NATO and ABCA Armies). Medical property accompanying patients of allied nations will be returned at once, if possible. If it is not possible, like items will be exchanged as in paragraph *a* above.

c. Medical Property of Coalition Forces or Allied Nations Without Ratified Standardization Agreements. Absence of a formal agreement, such as an Acquisition and Cross-Servicing Agreement, medical property accompanying patients of coalition and allied forces without ratified STANAGs will be returned to the parent nation as soon as practicable. Commanders should consult with their Staff Judge Advocate early in the planning process to ensure appropriate policy and procedures are developed and disseminated.

Section VI. TREATMENT TEAMS, MEDICAL SECTIONS, AND SPECIAL PURPOSE MEDICAL PLATOONS (ARMY OF EXCELLENCE/ AIRLAND BATTLE)

2-33. Combat Support Unit and Division Headquarters Treatment Team, L-Edition TOE

Treatment teams are organic to CS units and the division headquarters. With the exception of the combat engineer battalion, a medical support element in the light division normally consists of one treatment team. This treatment team is designed to provide Echelon I CHS for personnel of supported units. A treatment team normally with two ambulance teams is relatively small in comparison to a medical platoon; therefore, it will require augmentation from a supporting medical company in mass casualty situations.

2-34. Medical Section, Headquarters and Headquarters Battery, Division Artillery, L-Edition TOE

a. Organizations and Functions. The DIVARTY medical element includes a treatment team and an ambulance team. It is organized as shown in Figure 2-13. Personnel staffing of this section includes a PA, health care SGT, two health care specialists and an ambulance team.

(1) *Division artillery physician assistant*. The PA is the medical advisor to the DIVARTY commander and his staff. The PA works under the supervision of a physician, normally the division surgeon or treatment platoon leader of the MSMC. He coordinates with the division and brigade surgeons to ensure that all PAs/medical element leaders in the subordinate FA battalions, are working under the clinical supervision of a physician. Certain situations may require that the clinical supervision of PAs in FA units be passed to the physician in charge of the nearest supporting MTF. Such requirements, however, are coordinated through the division surgeon. The PA is responsible to the supervising physician for the medical treatment provided by DIVARTY medical personnel (inclusive of medical personnel assigned to FA battalions). His duties include—

- Operating the DIVARTY aid station.
- Planning and directing unit Echelon I CHS for members of the DIVARTY headquarters and FA battalions.
- Arranging for Echelon II CHS.
- Arranging for patient evacuation to the supporting medical company.
- Supervising the administration and maintenance of equipment, the supply function, technical training, and the employment of medical personnel.
- Examining, diagnosing, and treating (or prescribing courses of treatment for) patients, to include ATM for the trauma patient under the clinical supervision of a physician.
- Coordinating patient evacuation.

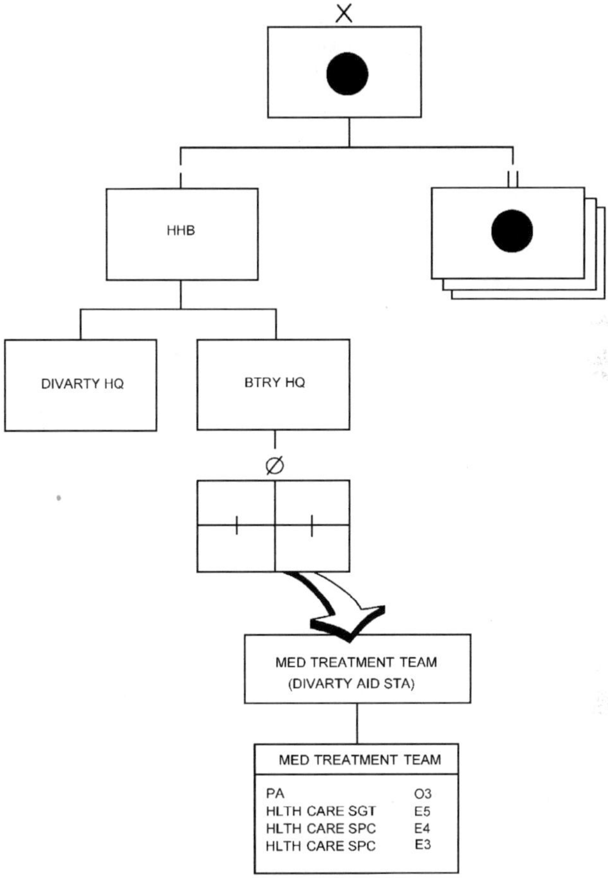

NOTE: THE LIGHT INFANTRY AND AIRBORNE DIVISION ARTILLERY DO NOT HAVE AN AMBULANCE TEAM.

Figure 2-13. Medical section, headquarters and headquarters battery, division artillery.

(2) *Health care sergeant.* The health care SGT assists the PA in accomplishing his duties; he supervises personnel on the treatment and ambulance teams. He prepares reports, requests general and medical supplies, maintains supply economy procedures, and maintains the ASL of expendable supplies. This NCO also performs triage and assists with ATM procedures in the care of trauma and NBC-insulted patients, and care and management of BF patients. He also performs routine patient care and NBC detection procedures. His duties also include—

- Establishing and assisting with operating the DIVARTY aid station.
- Maintaining the patient accountability/casualty reporting system.
- Maintaining MES.
- Conducting tactical and technical proficiency training for subordinate members of the section.
- Conducting sanitation inspections of troop living areas, food service areas, waste disposal areas, and potable water distribution points and equipment.
- Maintaining field health records of all patients seen according to AR 40-66.

(3) *Health care specialists.* These specialists assist the health care SGT in accomplishing his duties. They perform triage and EMT. Their specific duties include—

- Erecting and breaking down field medical shelter systems, to include chemical/ biological protective shelters.
- Performing patient care.
- Initiating patient records (FMC).
- Maintaining the patient daily disposition log.
- Operating and maintaining assigned vehicle, tactical radio, and power generation equipment. (Also may serve as a member on the battery field sanitation team.)

b. *Employment.* The medical section establishes an aid station near the DIVARTY headquarters and provides Echelon I CHS for members of the DIVARTY headquarters and headquarters battery (HHB).

(1) The section employs a treatment HMMWV, a cargo trailer, and two MESs (one trauma treatment set and one general sick call set).

(2) For communications, the section employs a telephone set (TA 312/PT) and is deployed in the HHB wire net. It employs an FM tactical radio and is deployed as designated by the DIVARTY SOI. This section also has access to the supporting medical company's tactical operations net to request Echelon II CHS.

c. Operations. The preceding paragraphs describe BAS operations; these are equally applicable to the DIVARTY BAS. Figures 2-8 and 2-9 show suggested layouts of a BAS.

d. Medical Evacuation. The DIVARTY HHB, depending on the type, may have one ambulance team to provide medical evacuation support. Those units assigned to the DIVARTY without an ambulance team are dependent on the supporting medical company. Evacuation of patients to and from the DIVARTY aid station is provided by the MSMC in the DSA.

e. Class VIII Supplies. The medical section maintains a 2-day (48-hour) stock level of Class VIII supplies for the HHB. Routine requests for medical supplies are submitted through command channels to the DMSO that is assigned to the MSMC. Class VIII supplies may be picked up by the requesting unit or forwarded to the DIVARTY BAS during routine ambulance runs. For emergency resupply procedures, see paragraph 2-25.

f. Property Exchange. See paragraph 2-32.

2-35. Treatment Team, Headquarters and Headquarters Battery, Field Artillery Battalion, L-Edition TOE

Organic to the HHB of the FA battalion is a treatment team, an ambulance team, and a combat medic section. The treatment team operates the BAS and the ambulance team provides limited ground ambulance medical evacuation support for the battalion. Trauma specialist from the combat medic section, are deployed with each firing platoon and with the service battery. Medical elements of the HHB, FA battalion are organized as shown in Figure 2-14. Personnel staffing for the treatment team include a PA, a health care SGT, two health care specialists, MOS 91W20, and one health care specialist, MOS 91W10.

a. Physician Assistant. In the absence of a physician, the PA is the principal advisor to the battalion commander and his staff in the areas of health and medical readiness. Working under the clinical supervision of a physician, he is the primary medical care provider for the battalion and supervises all activities of battalion medical personnel. The PA is trained in ATM procedures and as stated, works under the clinical supervision of a medical officer. He is responsible to the supervising physician for all treatment provided by medical personnel of the section. His specific duties include—

- Establishing and operating the BAS.
- Planning and supervising Echelon I CHS and coordinating with the supporting medical company for Echelon II CHS for the battalion.
- Treating, within his ability, patients reporting to him.
- Referring patients who require treatment beyond his capability to the supervising physician.
- Providing initial resuscitation (ATM) for the wounded.
- Training medical personnel and CLSs in emergency medical procedures.

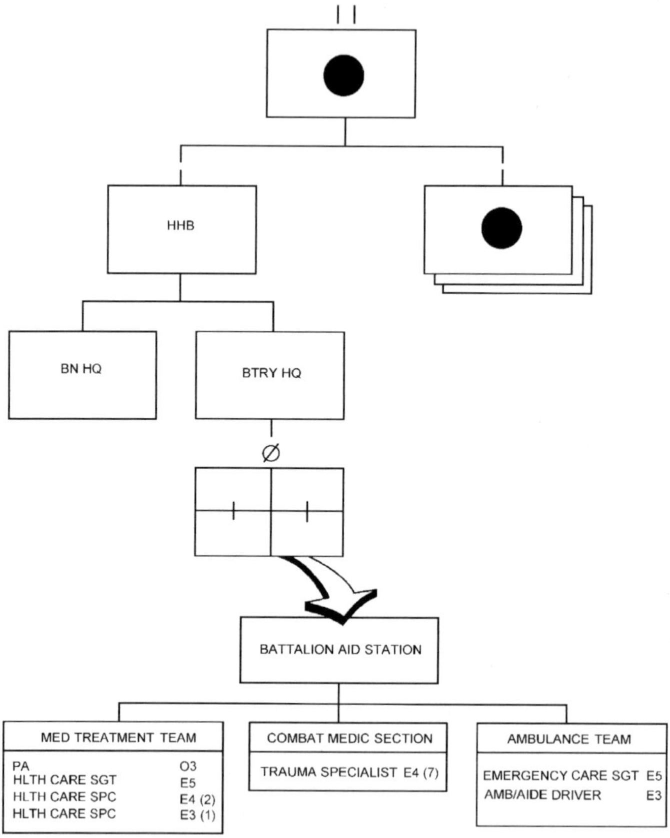

Figure 2-14. Medical elements, headquarters and headquarters battery, field artillery battalion.

b. Health Care Sergeant. This NCO assists the PA in accomplishing his duties. The specific duties of this NCO are the same as those described for the health care SGT in the DIVARTY HHB (refer to paragraph 2-34*a*[2]).

c. Health Care Specialists. The duties and functions of these specialists are the same as those discussed in paragraph 2-34*a*(3).

d. Trauma Specialists. Trauma specialists are allocated to a DS FA battalion on the basis of one to each firing platoon and the service battery. The duties and functions of trauma specialists are described in paragraph 2-20.

e. Employment. The treatment team establishes a BAS near the battalion headquarters and provides Echelon I CHS.

(1) The section employs a treatment HMMWV, a cargo trailer, and two MESs (one trauma treatment set and one general sick call set).

(2) For communications, the section employs a telephone set (TA 312/PT) and is deployed in the HHB wire communications net. It also employs an FM tactical radio and is deployed in the net designated by the battalion SOI. This section also has access to the supporting medical company's tactical operations net to request Echelon II CHS.

f. Operations. Paragraphs 2-19 describes a BAS operation; these are equally applicable to the FA BAS. Figures 2-8 and 2-9 show suggested layouts of a BAS.

g. Medical Evacuation. The HHB ambulance team evacuates patients to the BAS and to the supporting medical company if Echelon II ground ambulance support is not available.

h. Property Exchange. See paragraph 2-32.

2-36. Medical Treatment Team, Headquarters and Headquarters Company, Division Aviation Brigade/Combat Aviation Squadron, L-Edition TOE

a. Organization and Functions. The division aviation brigade/combat aviation squadron medical treatment team is organized as shown in Figure 2-15. Personnel staffing for this section include a flight surgeon, a health care SGT, and two health care specialists. The flight surgeon (brigade surgeon) is the medical advisor to the aviation brigade commander and his staff. He is the primary care physician of the brigade. The flight surgeon is responsible for treatment provided by the medical treatment team (brigade aid station). His duties include—

- Operating the brigade aid station.

- Examining and determining the medical qualification for flying status of aviators within the brigade headquarters, or aviators referred to him by units without a flight surgeon.

- Planning and directing Echelon I CHS for members of the brigade headquarters.
- Coordinating for evacuation of patients to the division clearing station.
- Coordinating for division CHS augmentation, as required..
- Supervising the administration and maintenance of equipment, the supply function, technical training, and the employment of medical personnel.
- Examining, diagnosing, treating, and prescribing courses of treatment for patients, to include ATM for trauma patients.

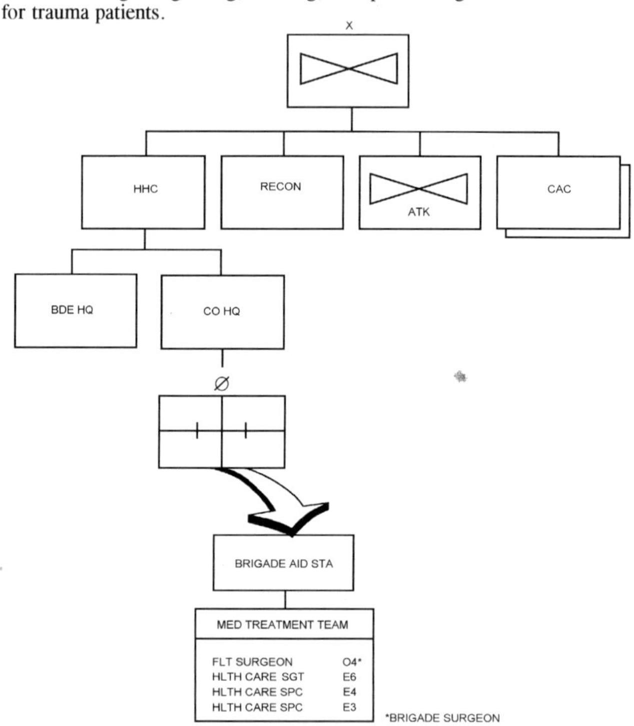

Figure 2-15. Medical treatment team, headquarters and headquarters company, division aviation brigade.

b. Operations. Paragraph 2-19 describes aid station operations; these are equally applicable to the DIVARTY BAS. Figures 2-8 and 2-9 show suggested layouts of a BAS.

c. Medical Evacuation. The brigade HHC medical section has no MEDEVAC assets. Evacuation of patients is provided by the supporting medical company.

d. Class VIII Resupply. See paragraph 2-25.

e. Property Exchange. See paragraph 2-32.

2-37. Medical Section, Headquarters and Headquarters Company, Attack Helicopter Battalion, Division Aviation Brigade, L-Edition TOE

a. Organization and Functions. The attack helicopter battalion medical section is organized as shown in Figure 2-16. Personnel staffing this section include a flight surgeon, a health care SGT, and two health care specialists. For further explanation, see paragraph 2-36*a*.

b. Property Exchange. See paragraph 2-32.

2-38. Medical Platoon, Headquarters and Headquarters Troop, Reconnaissance Squadron, Division Aviation Brigade, L-Edition TOE

a. Organization and Functions. The headquarters and headquarters troop (HHT) RECON squadron, division aviation brigade medical section is organized as shown in Figure 2-17. The medical section has a medical treatment squad, an ambulance squad, and a combat medic element.

(1) The patient treatment squad includes a flight surgeon (AOC 61N00), a PA, a section SGT/health care SGT (MOS 91W30), two health care SGTs (MOS 91W20), and three health care specialists (MOS 91W10).

(*a*) For flight surgeon responsibilities, see paragraph 2-36.

(*b*) The PA assists the flight surgeon in performance of his duties. He serves as the aviation brigade flight surgeon in the absence of the flight surgeon. His duties include—

- Examining and determining the medical qualification for flying status of aviators within the brigade headquarters; or aviators referred to his treatment section by units without a flight surgeon.

- Examining, diagnosing, treating, and prescribing courses of treatment for patients, to include ATM for trauma patients.

- Performing general technical health care and administrative duties (refer to paragraph 2-17*d*).

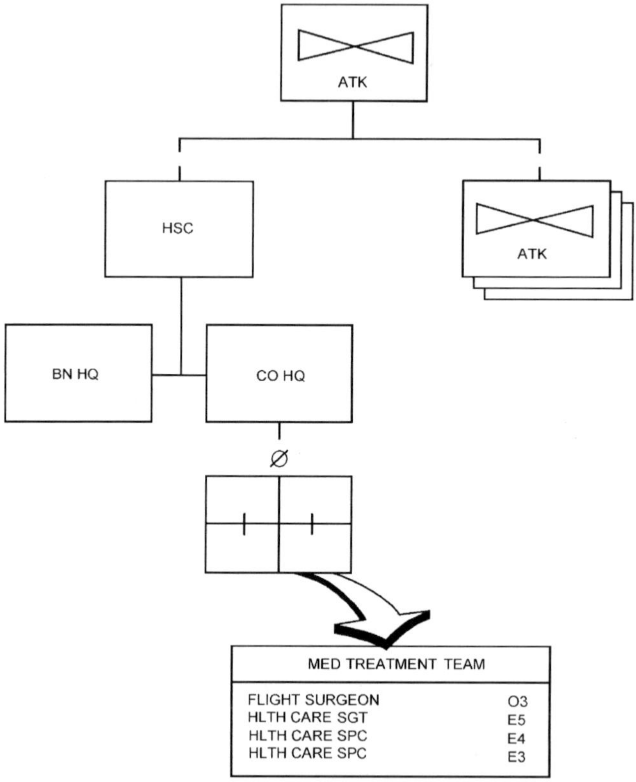

Figure 2-16. Medical section, headquarters and headquarters company, attack helicopter battalion.

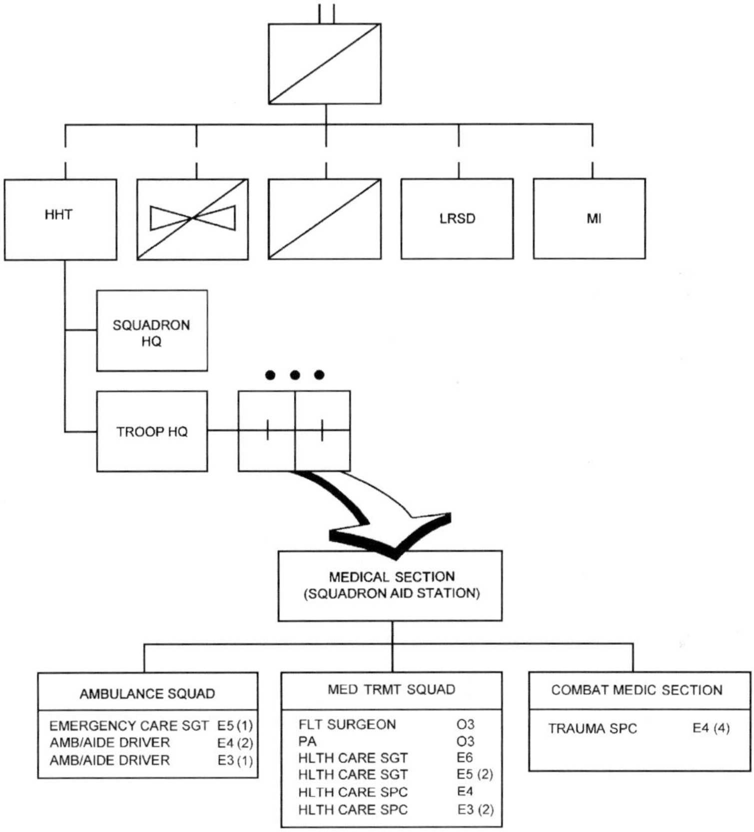

Figure 2-17. Medical platoon, headquarters and headquarters troop reconnaissance squadron.

(2) The ambulance squad has one emergency care SGT (MOS 91W20) and three ambulance/ aide drivers (MOS 91W10).

(3) The combat medic element has one trauma specialist, MOS 91W10.

b. Section Sergeant. This NCO, also a health care SGT, assists the flight surgeon in accomplishing his duties. The specific duties of this NCO are the same as those described for the health care SGT in the DIVARTY HHB (refer to paragraph 2-34).

c. Health Care Specialists. The duties and functions of these specialists are the same as those discussed in paragraph 2-34.

d. Trauma Specialists. The duties and functions of trauma specialists are described in paragraph 2-20.

e. Ambulance Squad. Paragraph 2-21 describes duties of ambulance squad members.

f. Employment. The medical section establishes a BAS near the squadron headquarters and provides Echelon I CHS for members of the squadron.

(1) The section employs a treatment HMMWV, a cargo trailer, and two MESs (one trauma treatment set and one general sick call set).

(2) For communications, the section employs a telephone set (TA 312/PT) and is deployed in the headquarters and headquarters support company's wire communications net. It also employs an FM tactical radio and is deployed in the net designated by the squadron SOI. This section has access to the supporting medical company's tactical operations net for requesting Echelon II CHS.

g. Operations. Paragraphs 2-19 describes a BAS operation; these are equally applicable to the squadron BAS. Figures 2-8 and 2-9 show suggested layouts of a BAS.

h. Medical Evacuation. Evacuation of patients from the squadron aid station is provided by the supporting medical company.

i. Medical Supply. The medical section maintains a 2-day (48-hour) stockage level of medical supplies for the squadron. Routine requests for Class VIII resupply are submitted through command channels to the DMSO. Supplies may be picked up by the requesting unit or forwarded to the BAS during routine ambulance runs. For emergency resupply procedures, see paragraph 2-25.

j. Property Exchange. See paragraph 2-32.

2-39. Medical Section, Headquarters and Headquarters Company, Division Headquarters, L-Edition TOE

a. Organizations and Functions. The HHC division headquarters medical treatment team is organized as shown in Figure 2-18. Personnel staffing of this section includes a PA, a health care SGT, and two health care specialists.

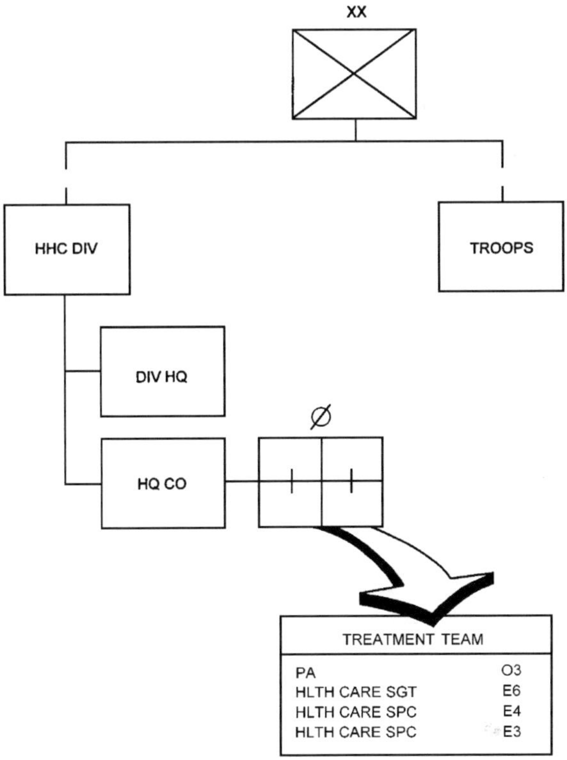

Figure 2-18. Medical section, headquarters and headquarters company, division headquarters.

(1) *Physician assistant.* The PA is responsible for the medical treatment provided by HHC medical personnel. He works under the clinical supervision of the division surgeon. In the division treatment team, the PA is the principal advisor to the division surgeon in the areas of PA affairs, executive medicine issues, and quality assurance/implementation. The specific duties of the PA are the same as those described in the DIVARTY HHB (refer to paragraph 2-34).

(2) *Health care sergeant.* Refer to paragraph 2-34.

(3) *Health care specialists.* Refer to paragraph 2-34.

b. Employment. The medical treatment team establishes an aid station near the division headquarters and provides Echelon I CHS for members of the division HHC.

(1) The section employs a treatment HMMWV, a cargo trailer, and two MESs (one trauma treatment set and one general sick call set).

(2) For communications, the section employs a telephone set (TA 312/PT) and is deployed in the HHC wire communications net. It also employs a FM tactical radio and is deployed in the net designated by the division SOI. This section has access to the MSMC's tactical operations net to request Echelon II CHS as required.

2-40. Combat Medic Section, Headquarters and Headquarters Company, Combat Engineer Battalion, L-Edition TOE

a. Organization and Functions. The combat medic section of the combat engineer battalion is organized as shown in Figure 2-19. Personnel staffing of this section includes a section SGT and ten trauma specialists. The combat medic section provides EMT and treatment of minor wounds and injuries. It coordinates for and/or requests MEDEVAC support as required.

b. Section Sergeant. The section SGT, MOS 91W30, prepares reports, requests general and medical supplies, maintains supply economy procedures, and maintains the ASL of expendable supplies. He supervises combat medic section personnel. He coordinates Echelon I CHS from supported maneuver battalion medical platoon and Echelon II CHS from the supporting medical companies. His duties also include—

- Assigning tasks to trauma specialists.
- Providing and/or coordinating for Class VIII resupply for trauma specialists when deployed with engineer platoon or squad.
- Conducting tactical and technical proficiency training for subordinate members of the section.
- Conducting sanitation inspections of troop living areas, food service areas, waste disposal areas, and potable water distribution points and equipment.
- Coordinating and conducting CLS training for the battalion.
- Providing medical planning input to the S1 on battalion operations.

c. Trauma Specialists. The duties and functions of trauma specialists are described in paragraph 2-20.

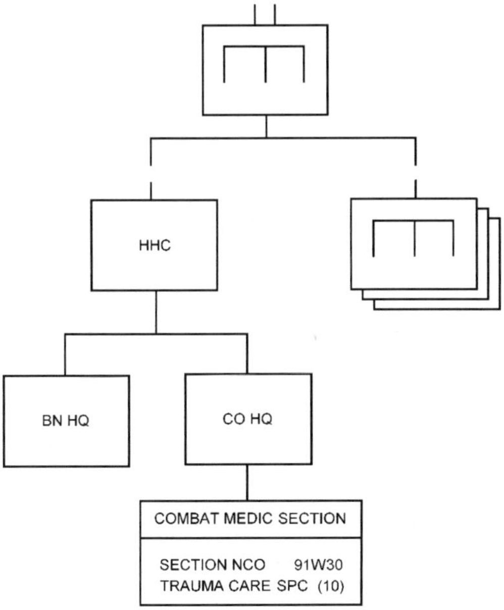

Figure 2-19. Medical platoon, headquarters and headquarters company, engineer battalion.

Section VII. ADDITIONAL MEDICAL ASSETS OPERATING IN THE BRIGADE AREA OF OPERATIONS (FORCE XXI/DIGITIZED DIVISION)

2-41. Treatment Squads/Teams from the Forward Support Medical Company, F-Edition TOE

The treatment squad provides emergency and routine sick call treatment to soldiers assigned to supported units. These teams can perform their functions while located in the FSMC area, or can operate independently of the FSMC for limited periods of time. The squad has the capability to split and operate as separate treatment teams (Teams Alpha and Bravo) for limited periods of time. While operating in these separate modes, they may operate two separate treatment stations. Normally, a squad or team deploys forward to augment or reinforce maneuver battalion medical platoons. Ambulance squads/teams may be deployed to AXP especially when there are extended evacuation routes. It can be assigned to reinforce or reconstitute battle losses of maneuver battalion medical platoons.

2-42. Forward Surgical Team, A-Edition TOE (Force XXI/Medical Reengineering Initiative)

Corps-level initial surgical support will be provided by the FST. The FST (Corps), TOE 08518LA00, and the FST (Airborne/Air Assault Division/ACR [Light]), TOE 08518LB00, are clinically standardized modules regardless of their assignment. These teams are comprised of 20 personnel and each has two operating room (OR) tables. The FST is organized into four functional areas—triage-trauma management, surgery, recovery, and administrative/operations. The mission of the FST is to provide a rapidly deployable immediate surgical capability enabling patients to withstand further evacuation. The requirement to project surgery forward increases as a result of the extended battlefield. This small, lightweight surgical team is designated to provide surgical augmentation to the FSMCs in support of the maneuver brigades, brigade TF, or the Interim Brigade Combat Team (IBCT). The FST is capable of continuous operations with divisional or nondivisional medical companies/troops for up to 72 hours; the ability to continue operations is limited by personnel fatigue/exhaustion and available supplies. The FST provides urgent, initial surgery for otherwise nontransportable patients. The FST's surgical capability is based on two OR tables with a surgical capacity of 24 OR table hours per day. Other capabilities include—

- Emergency medical treatment, to include assets to receive, triage, and prepare incoming patients for surgery.

- Surgery, including initial surgery and continued postoperative care for up to 30 critically wounded or injured patients over a period of 72 hours with the FST's organic MESs prior to resupply.

- Nursing care. Postoperative acute nursing care for up to eight patients, simultaneously, prior to further medical evacuation.

- Rapid strategic deployability. The team's personnel and equipment (less vehicles) are capable of deploying in one C-130 aircraft for initial entry missions, when required. The FST is capable of subsequent movement by helicopter sling-load operations.

- Tactical mobility. The team is 100 percent mobile with organic vehicles; it has a total of six HMMWVs.

For definitive information of the FST, see FM 8-10-25.

2-43. Forward Support Medical Evacuation Team, L-Edition TOE (Force XXI/Medical Reengineering Initiative)

The brigade may be augmented with a corps FSMT. When deployed forward to the BSA, the FSMT leader coordinates the air ambulance team's evacuation missions. The FSMT, assisted by the support operations section, provides real-time tactical information to the air ambulance crew about evacuation missions from the brigade combat team units/elements to supporting brigade MTFs. When air ambulances operate forward of the BSA, they will execute the A2C2 plan through the maneuver brigade S3. The FSB support operations section provides planning and coordination between aeromedical evacuation and the supported maneuver brigade. The brigade S3 provides the A2C2 plan that includes the air corridors, air control points, and communications checkpoints. The brigade S3 will provide updates as required. Air ambulances deployed to the BSA provide medical evacuation from forward areas (BAS) back to the BSA. Air ambulance evacuation from the point of injury will be METT-TC-dependent. Corps air ambulances providing GS evacuate from the BSA to supporting corps MTF. Aeromedical elements provide around the clock immediate response evacuation from either the BSA or their location based on METT-TC. To accomplish this, elements must maintain a close tie with the A2C2 system in the brigade. The brigade A2C2 element provides an airspace plan through the division OPORD/OPLAN A2C2 annex. The aircrew must also be familiar with the daily airspace control order and the airspace control plan. These documents contain all airspace control measures (ACM), to include free fire areas, no-fly fire areas, restricted operations zones, and established and standard Army aircraft flight routes. These routes and ACM change on a daily basis and cannot be integrated into the division OPORD. The BSS will ensure all A2C2 information is provided to corps aeromedical elements. The BSS does not generate A2C2 information, but does provide A2C2 planning information to division A2C2 elements. For definitive information on the corps air ambulance company and its FSMT that deploy forward into the brigades' AO, see FM 8-10-26.

2-44. Corps Ground Ambulance Company, Either the L-Edition TOE or the A-Edition TOE Supporting Force XXI/Medical Reengineering Initiative

The corps ground ambulance company is assigned to the corps MEDEVAC battalion. The basis of allocation within the combat zone is one per division supported. The current Army of Excellence ambulance company has four ambulance platoons with each platoon having 10 ambulances each. Under the MRI, the new MRI ground ambulance company will have a total of 24 ground ambulances. When deployed to the division, the ground ambulance company is attached to the MSB or division support battalion for Force XXI. The mission of the ground ambulance company in the division is to provide medical evacuation support to the division's maneuver brigades and to other divisional units and corps units operating in the division, as required. Normally, corps ground ambulances provide medical evacuation from the FSMC located in the BSA and from either the MSMC or the DSMC (Force XXI) to the supporting corps combat support hospital. The corps ground ambulance will reinforce the MEDEVAC assets in the medical companies of the division, as required.

2-45. Corps Combat Stress Control Augmentation, A-Edition TOE (Force XXI/Medical Reengineering Initiative)

The division may be augmented with additional CSC personnel, if requested. The base of allocation for the CSC medical detachment is one per division supported by the corps. The CSC medical detachment provides complete MH and combat stress preventive and treatment services in DS of division and corps personnel deployed forward. The new MRI detachment is a 43-person unit composed of a headquarters, a CSC preventive section and a CSC fitness section. The old Medical Force 2000 CSC medical detachment had 23 personnel and was designed to be a corps-level package to augment the organic MH sections of the divisions. Whereas, the new MRI CSC medical detachment retains the mission of providing DS to a division's maneuver brigades and general/reinforcing support to the DSA, including corps units in those areas. In addition, the detachment now augments area support in the corps immediately behind the division. The detachment must function with its elements widely dispersed, some working in and for the supported division and others working in the corps for the medical command/brigade. The CSC medical detachment personnel provide CSC planning, consultation, training, and staff advice to C2 headquarters and the units to which they are assigned/attached regarding—

- Combat and operational stressors affecting the troops.
- Mental readiness.
- Morale and cohesion.
- Potential for BF casualties.

The detachment and its personnel are dependent on units that they are attached for support, to include—

- Food service.
- Water distribution.
- Medical treatment.
- Logistical support including Class VIII items.
- Patient administration (detachment has one patient administration specialist, MOS 71G10, that works with the supporting unit).

For definitive information on the CSC medical detachment, Medical Force 2000 and MRI, see Change 1, FM 8-51.

CHAPTER 3

COMMAND AND CONTROL

Section I. PREPARATION FOR COMBAT HEALTH SUPPORT

3-1. Plans

a. Mission Analysis. Planning starts with mission analysis. The battalion begins mission analysis when the brigade provides a warning order (WARNO). The mission analysis is Step 1 of the military decision-making process (MDMP). See FM 101-5 for further discussion on the MDMP. For guidance on military decision making in abbreviated planning for a time-constrained situation, see FM 101-5. The battalion headquarters conducts concurrent planning with the brigade headquarters or after the brigade plan is developed. The battalion staff may receive additional information from the brigade staff elements to assist them with the planning process. This information is normally transmitted in a force text e-mail message via the tactical local area network (LAN). As part of the mission analysis and based on the battalion commander's intent and guidance, the medical platoon develops CHS estimates for supporting battalion operations. An understanding of the battalion's time lines or battle rhythm will assist the battalion medical platoon leader and field medical assistant in developing the CHS input, through the battalion S1, to the battalion's OPLAN/OPORD. The battalion surgeon and field medical assistant work with and through the battalion S1 for mission analysis input. See Chapter 5 and Appendix H of FM 101-5 for additional information on WARNOs. Mission analysis includes—

- Assessing CHS capabilities (organic and attached assets with current status and location).
- Assessing limitations (specify reason that CHS assets are not available).
- Identifying specified, implied, and essential CHS tasks in the brigade OPORD.

The following are examples of subject areas that should be addressed during mission analysis:

- Treatment (to include surgical requirements).
- Emergency dental treatment.
- Combat stress control.
- Preventive medicine.
- Medical evacuation support by air and ground ambulances (and nonmedical evacuation platforms, if necessary).
- Class VIII resupply.
- Medical maintenance.
- Nuclear, biological, and chemical operations.

- Threat to treatment and evacuation assets that is capable of causing CHS failure.

- Casualty estimates (number and types of casualties).

- Terrain effects on location, acquisition, and evacuation of casualties.

- Current medical status of battalion personnel.

b. Battalion Course of Actions. Battalion COAs development/analysis and wargaming are accomplished after mission analysis. Course of action development and wargaming result in the production of the OPORD and the CHS annex. During wargaming, the evacuation and treatment facets of the medical plan are synchronized with the overall battalion plan. The S1 will provide the overall casualty figures by battalion and, possibly, company. The questions of how many casualties, at what point in the fight (when), where they will occur, and how they are produced (direct fire, artillery, chemical, and so forth) can be forecasted based on input from the S1. During the wargaming, the S2 will portray enemy capabilities and likely actions. The S3 will focus on friendly actions. The medical platoon leader needs to pay careful attention to this exchange. This will be the best predictor of what, when, where, how, who, and other information that will be useful in adjusting the CHS plan. This information allows the medical platoon leader and field medical assistant options to select preplanned locations for positioning ambulances or treatment teams. Locating the treatment teams in the appropriate locations is of the utmost importance. The published movement planning factors for inside the division AO for wheeled and tracked vehicles in good terrain and weather are 8 kilometers and return in 1 hour (or 16 kilometers per hour). See FM 8-55 for additional planning guidance. The trauma specialist's goal is to get the casualty to ATM within 30 minutes. For an ambulance to leave the BAS and pick up a patient and return within 30 minutes, it must be within 4 kilometers of the soldier's point of injury. Keep in mind that this is under favorable conditions. Limited visibility, difficult or unfamiliar terrain, obstacles (friendly and enemy), and enemy actions will make the evacuation mission longer. If the BAS is farther than 4 kilometers away, it starts out as an impossibility. The METT-TC will govern specific solutions. Supporting the fight and maintaining a good support distance becomes a definite challenge because most of your evacuation routes can potentially be covered by enemy weapons and direct and indirect fire. For example, Russian manufactured AT-5 spandrel antitank-guided missiles have a 4-kilometer direct fire range. The medical platoon plan must take this threat into account. Information obtained from the S2 and S3 will also provide the triggers that will allow medical elements to occupy these positions at appropriate times and in a manner that reduces the risk from threat actions. This information is incorporated into the CHS plan and is published in the battalion order. The following are examples of key areas that will be analyzed during this process; they include—

- Casualty estimates broken down to the lowest level possible, by TF, by phase line, and so forth.

- Battalion- or brigade-directed actions as part of the squadron or brigade CHS plan.

- Current medical platoon equipment status (maintenance status on all the platoon's key items of equipment, both medical and nonmedical).

3-2. Operation Order

Once the battalion receives the brigade WARNO, it begins mission analysis and determines its tactical plan. This process continues until the full brigade OPORD is issued. Part of the mission analysis is to assess the brigade CHS plan for its adequacy of support to the battalion. If the medical platoon leader finds a problem, he briefs this to the battalion commander. The mission analysis brief will be after the staff has analyzed the full OPORD. Another part of the analysis is determining the employment and emplacement of medical treatment elements (BAS/treatment team) based on the brigade plan. This information is provided to the battalion S3 and he updates the battalion plan. This is normally accomplished prior to the brigade combined arms rehearsal. When the battalion commander approves the OPLAN, it becomes the OPORD. The OPLAN and OPORD are developed by the S3 section using input from each of the staff elements with the S1 being the staff coordinating element for CHS. The battalion OPORD is revised or updated based on mission changes. Table 3-1 is an example of an OPORD/OPLAN outline format.

a. Matrix Operation Order. A matrix OPORD may be used as an alternative to the standard five-paragraph OPORD. The purpose of the matrix OPORD is to cut orders production time and to provide subordinates more time for RECON, preparation, and rehearsal. There is no standard format for a matrix OPORD. Matrix orders expand on the execution matrix found on many operations overlays. The single-page format may include all signal information for the day of the operation and it can be placed in the corner of a map case for easy reference. Matrix orders are usually issued with standard operations, intelligence, and fire support overlays. Rather than a five-paragraph order outline format as seen in Table 3-1, the medical platoon leader is more likely to see and work with a matrix OPORD. Figure 3-1 is an example of a matrix OPORD.

b. Medical Support Matrix. The medical support matrix should be integrated with the tactical overlay. Figure 3-2 is a sample format for a medical support matrix. If deviation from the matrix occurs, the BAS location must be known at all times. The BAS should remain on location as long as practical. Extra first aid medical supplies can be issued to maneuver elements for resupply of CLS.

Table 3-1. Sample Outline Format for an Operation Order/Plan

(CLASSIFICATION)

TIME ZONE USED THROUGHOUT THE PLAN (ORDER)

REFERENCES:

TASK ORGANIZATION:

1. **SITUATION**
   A. ENEMY FORCE
   B. FRIENDLY FORCE
   C. ATTACHMENT AND DETACHMENTS
   D. ASSUMPTION (OPLAN ONLY)

2. **MISSION**

3. **EXECUTION**
   INTENT:
   A. CONCEPT OF OPERATION
      (1) MANEUVERS
      (2) FIRES
      (3) RECONNAISSANCE AND SURVEILLANCE
      (4) INTELLIGENCE
      (5) ENGINEER
      (6) AIR DEFENSE
      (7) INFORMATION OPERATIONS
   B. TASK TO MANEUVER UNITS
      (1) ENGINEER
      (2) AIR DEFENSE
      (3) FIELD ARTILLERY
   C. TASKS TO COMBAT SUPPORT UNITS
      (1) INTELLIGENCE
      (2) FIRE SUPPORT
      (3) SIGNAL
      (4) NBC
      (5) PROVOST MARSHAL
      (6) PSYOP
      (7) CIVIL MILITARY
      (8) AS REQUIRED
   D. COORDINATING INSTRUCTIONS
      (1) TIME OR CONDITION WHEN A PLAN OR ORDER BECOMES EFFECTIVE
      (2) COMMANDER'S CRITICAL INFORMATION
      (3) RISK REDUCTION CONTROL MEASURES
      (4) RULES OF ENGAGEMENT
      (5) ENVIRONMENTAL CONSIDERATIONS
      (6) FORCE PROTECTION AS REQUIRED
      (7) AS REQUIRED

4. **SERVICE SUPPORT**
   A. SUPPORT CONCEPT
   B. MATERIEL AND SERVICE
   C. MEDICAL EVACUATION AND HOSPITALIZATION
   D. PERSONNEL
   E. CIVIL MILITARY, AS REQUIRED
   F. AS REQUIRED

5. **COMMAND AND SIGNAL**
   A. COMMAND
   B. SIGNAL

ACKNOWLEDGE:
NAME (COMMANDER'S LAST NAME)
RANK (COMMANDER'S RANK)
OFFICIAL: NAME AND POSITION

ANNEXES

(CLASSIFICATION)

SEE FM 101-5 FOR DEFINITIVE INFORMATION ON OPERATION PLANS/ORDERS. ADDITIONAL INFORMATION IS PROVIDED IN APPENDIX 3 (PERSONNEL) OF ANNEX I, OR IN A SEPARATE APPENDIX FOR CHS.

TF 2-69 ARMOR DEFENSIVE MATRIX OPORD #1-88 DTG 011000A JAN 88 COPY 21 of 25 COPIES

IF CALL SIGN ____ (42.10) BN MORTAR/FS NET (52.40) BDE FS NET (38.55) CDR 16 S3 91 FSO 07 MAIN 76 XO 34 S3A 23 FAC 37 PLDR 12 PSG 15 AJ1 (48.75) S2 39 S4 AA BHO 22 ADA V1T15 (42.00) AJ2 (45.00)	ATTACHMENTS/DETACHMENTS ~~A/1 58~~ ~~72 ENG~~

UNIT/STATUS	TM MECH G	TM TANK G	Co C G	Co D G	SCOUTS G	T2 ENGR G	MORTARS G
CALL SIGN/FREQ	IIM (40.25)	V6D (52.65)	NBY (50.10)	M9U (512.25)	D41 (142.89)	T5V (52.50)	ABE (52.40)
TASK ORG	1 2/8 2 2 1	1 3 3/4 2	1 2 3 3	I I I 1 2 4	EYES EARS 3/D 5	1 3	BLUE { FDC RED { FDC

IF MISSION: TF2-69 defends in sector 020600Jan 88 from coords to coords and coords to coords to destroy MRR.

CDRS INTENT: Occupy deception positions along PL DOG; NLT 020400 move to prepared positions along PL CAT; destroy en in EA LION (Ave 1) or EA TIGER (Ave 2) allowing no penetration of PL MOUSE.

(5) SP MOVE ORDER (3)	6	011400 / 2	4	7	011000 / 1	3	PRI TO TM TANK / 5
(1) OCCUPY PL DOG (DECEPTION) ORIENT (9)	BP 2	BP 3	BP 1	BP 4	SCREEN IF FRONT ALONG PL BEAR	PREPARE OBSTACLES ALONG PL CAT	(COORDS) / PRI INIT TO SCOUTS
(1) PL CAT (BALANCED ATK) (6)	PB6 / EA LION	BP 7 / EA TIGER	BP 5 / BA LION	BP 8 / EA TIGER		CONTINUE OBSTACLE & SURVIVABILITY POSITION WORK	NU213562 / PRI TM MECH when EN crosses PL DOG
(4) MAIN ATK Ave 1 (2)		o/o BP 11 / Ave 1		BP 9 / EA LION	Screen if right flank	PRI effort to BP 11	PRI TM MECH
(8) MAIN ATK Ave 2 (10)	o/o BP 12 / Ave 2	BP 7 / EA TIGER	BP 10 / EA TIGER	BP 8 / EA TIGER	Screen if left flank	PRI effort to BP 12	PRI TM TANK
FIRES 2 to FA PRI to 2-69 ·PRI Init sets ·FPFs: ARTY: TM Mech MORT: TM Tank Illum planned in AB 4004, 5, 6, 7 but o/o TF CDR FASCAM planned in AB 4004#5 o/o Bde Cdr	DECEPTION ·OCC BP 6 ·RECON BP 12 ·Conserve MIN ·Fire IGTS: AB 4004 AB 4006 ·close mine field AB 4006 after secure pass	SKYLINE VC ·OCC BP 7 ·RECON BP 11 ·OBSV MA 12 ·Fire TGTS AB 4005 AB 4007	OF TANKS ON ·OCC BP 5 ·Prep BP 10 ·OBSV NA 11 TO BP 10 at night; RPT Time	·OCC BP 8 ·Prep BP 9 ·OBSV NA 122 ·3/D ATNS 020500	·Clear TF Section NLT 011400 ·3/D OPCON until 020500 ·Fire Tgts AB 4002 AB 4003 ·Observe NAI 12		DISPLACE BY SECTION
OBSTACLES	Minefield AB 4006	Minefield AB 4007	Wire Obstl #1	Wire Obstl #2		Minefield 1-AB 4006 2-AB 4007	
FIGHTING POSN PRI (No.) LOC	#1(4) BP 6	#2(10) BP 7	#3(8) BP 5 #5(4) BP 10	#4(8) BP 8 #6(4) BP 9		#7(6) BP 5 #8(6) BP 8	
CDR BP 6	S3 BP 7	MAIN CP CP 2	CBT TNS CP 4	UMCP CP 6	FLD TNS BSA (CP 10)	LRP A CP 12	LRP B CP 8
ADA Yellow Tight	MOPP 2	OEG 50 cGy	SOI KTV 1062T	PYRO Main Atk AV 01 Red Star AV 02	AJ CODEWORD Sprint	PIR (1) EN (2) ALOC	STAND TO 0600 Daily

Figure 3-1. Matrix operation order.

UNITS	CROSS LD	CROSS PL RED	CROSS PL BLUE	ON OBJ DALLAS
CO A AND CO B (-)	TT 1 (NB 583492) (TOM)	TT 1 (NB 585501) (JIM)	TT 1 (NB 587507 (BILL)	TT 1 (NB 591510) (GREG)
CO C AND RESERVE	TT 2 (NB 581489) (VICKY)	TT 2 (NB 583499) (BETH)	TT 2 (NB 585505) (JANET)	TT 2 (NB 589508) (JOY)
RM 1: USE ROUTE STEVEN				
RM 2: USE ROUTE ANNA				

Figure 3-2. Medical support matrix.

c. Implementation of Combat Health Support Annex. For successful implementation of the CHS annex of the battalion plan, the CHS plan must be coordinated and synchronized with the battalion plan so that CHS requirements are met. The medical platoon leader may receive additional coordinating instructions from the BSS as the CHS annex is developed. This additional information permits the medical platoon leader to make informed decisions as he develops the CHS annex and affords some additional time for better development of good initial plans. For successful implementation of the CHS annex of the battalion's/ squadron's plan, the CHS plan must be coordinated and synchronized with the overall plan. To achieve optimal synchronization, the battalion CHS plan is rehearsed as an integral part of the combined arms plan at the combined arms rehearsal. The CHS rehearsal by itself is a technique that will increase understanding and synchronization, but is not as effective as when it is integrated into the combined arms rehearsal. The rehearsal of the CHS plan will allow battalion medical elements and leaders to analyze the tactical CHS plan to ascertain its feasibility, its common sense, and the adequacy of its C2 measures prior to execution.

3-3. Rehearsal

To achieve optimal synchronization, the battalion CHS plan is rehearsed as an integral part of the combined arms plan at the combined arms rehearsal. Medical platoon leaders and field medical assistants provide input to the battalion plan and develop the concept for the battalion CHS plan. During the decision-making/ orders process, they identify critical events and synchronize the CHS plans. In addition to medical locations on the CSS overlay, these plans indicate the triggers for CHS events. At the battalion rehearsal, battalion leaders practice their synchronized plans that include CHS. The sequence of events for the CHS portion of the battalion rehearsal includes—

- Ambulance teams practicing execution of triggers for area medical support responsibilities and triggers for movement of supporting FSMC ambulances.

- The medical platoon SGT detailing the concept and procedures for MEDEVAC (both ground and air ambulances) in the battalion.

- The battalion XO or S4 explaining triggers for BAS/treatment team movement and ensuring that the battalion HHC and maneuver companies understand when and where the BAS/treatment teams are located.

- The medical elements providing projected triggers and times they will be at projected locations.

The CSS/CHS annex of the battalion OPORD that includes map overlays is the conclusion of the medical planning efforts; the rehearsal is the culmination of the preparation phase for an operation. The medical platoon leader has the responsibility for rehearsing CHS operations. Rehearsals are done to achieve a common understanding and a picture of how the plan will be implemented.

- All plans must be completed prior to the battalion rehearsal.

- The CHS portion of the battalion rehearsals should focus on the events that are critical to mission accomplishment. A successful rehearsal ensures explicit understanding by subordinate medical personnel of their individual missions; how their missions relate to each other; and how each mission relates to the commander's plan. It is important for all medical echelons to see the total CHS concept.

- Rehearsing key CHS actions allows participants to become familiar with the operation and to visualize the "triggers" which identify the circumstances and timing for friendly actions. This visual impression helps them understand both their environment and their relationship to other units during the operation. The repetition of critical medical tasks during the rehearsal helps leaders remember the sequence of key actions within the operation and when they are executed.

- The battalion OPORD is then issued through effective troop leading procedures.

Section II. TROOP-LEADING PROCEDURES

3-4. Eight Steps of Troop Leading

The commander makes most tactical decisions. He then announces them in the form of orders that include his intent and concept of the operation. Based on these orders, the medical platoon leader uses troop-leading procedures to organize his time during planning and preparation and to translate the operation into instructions his soldiers can understand. He can then lead the platoon more effectively in the execution of the mission. Troop leading is a dynamic process that begins when the unit receives a new mission or is notified by a WARNO that a new mission is imminent. Whenever possible, troop-leading procedures are integrated and accomplished concurrently rather than sequentially. Time management is the key. The medical platoon leader normally uses one-third of the available time to plan, prepare, and issue the order; his field medical assistant, platoon SGT, ambulance squad leaders, and treatment team leaders then have the remaining two-thirds of the time available to prepare their ambulances and MES to support the operation. The following discussion focuses on the eight steps of troop leading procedures:

- Receive and analyze the mission.

- Issue a WARNO.
- Make a tentative plan.
- Initiate movement.
- Conduct RECON and coordination.
- Complete the plan.
- Issue the order.
- Supervise and refine.

a. Receive and Analyze the Mission. The medical platoon leader normally receives his orders as an oral OPORD or as a fragmentary order (FRAGO) updating a previously issued OPORD. Graphics may be copied from the S3's overlay or sent by digital transmission. Initial coordination within the battalion S1, medical platoon personnel, and with the supporting FSMC should be accomplished upon receipt of the mission.

NOTE

Before the OPORD or FRAGO arrives, the medical platoon leader may receive a series of WARNOs from the battalion S3 providing advance notice of an impending operation. The medical platoon leader should disseminate all pertinent information contained in the WARNO as quickly as possible after they are received. The battalion S4 may be the one who issues orders to the medical platoon since he has tactical control of the combat trains.

(1) Upon receipt of the WARNO, the medical platoon leader's first task is to extract the CHS mission based on the battalion commander's guidance and intent. The key to understanding the medical platoon CHS mission, as part of the battalion team, lies in two elements of the plan—the commander's intent and the purpose he envisions for the battalion and each company.

(2) The medical platoon leader's knowledge of the intent and purpose allows him to use his initiative and to be proactive and exploit battlefield opportunities to accomplish the CHS mission. If the medical platoon leader does not understand the intent or purpose, he should ask the commander for clarification.

(3) The medical platoon leader analyzes the mission using the factors of METT-TC. These factors allow the platoon leader to identify the platoon's purpose; the specified, implied, and essential tasks it must perform; and the time line by which the platoon will accomplish those tasks. The following outline

of METT-TC factors will assist the medical platoon leader in analyzing the mission and creating a time line. The medical platoon leaders need the answers to the questions pertaining to METT-TC.

(*a*) *Mission.*

- What is the battalion commander's intent?

- What are the current capabilities (organic and attached assets with current status and locations)?

- What are the specified, implied, and essential CHS tasks in the battalion and brigade OPORD?

- What are the limitations (CHS assets that are not available, specify reason)?

- What other tasks must be accomplished to ensure mission success (implied tasks)? Implied tasks are those that are not specified in the OPORD but that must be done to complete the mission. They do not include tasks that are covered in the unit SOP. The medical platoon leader identifies implied tasks by analyzing the enemy, the terrain, friendly troops available, and the operational graphics.

- What is the current patient status (for example, awaiting evacuation)?

- Are patient evacuation vehicles required to use contaminated routes?

- Are patient decontamination stations required?

- Where are the locations of the treatment team and of BAS or other MTF providing area medical support? (Current/projected?)

- What are the area medical support responsibilities?

(*b*) *Enemy.*

- What have been the enemy's recent activities?

- What is the composition of the enemy's forces?

- What are the capabilities of his weapons?

- What is the location of current and probable enemy positions?

- What is the enemy's most probable COA?

Enemy information is included in paragraph 1 of the OPORD. It is important that the medical platoon leader analyze this information in terms of how the medical platoon supports the operation. For example, the threat the enemy imposes on the battalion will influence how CHS operations are conducted based on the projected number of casualties the enemy will inflict on the battalion.

(*c*) *Terrain (and weather).* The medical platoon leader analyzes the terrain using the factors of OAKOC (obstacles; avenues of approach; key terrain; observation and fields of fire; and cover and concealment). Elements of this analysis include the following:

- *Obstacles.*
  - Where are natural and existing obstacles located, and how can they affect medical treatment and evacuation?
  - Where are likely areas for enemy-emplaced obstacles, and how can they affect maneuver?
  - Are there bypasses, or must obstacles be breached?
- *Avenues of approach.* Where are the best avenues of approach (mounted and dismounted) for enemy and friendly forces? These are considerations for determining evacuation routes and in planning for future locations for the BAS.
- *Key terrain.*
  - Where is the key terrain? Will FM communications be affected?
  - How can key terrain be used to support the mission?
- *Observation and fields of fire.*
  - Are these influenced by the key terrain that dominates avenues of approach?
  - Where can the enemy observe and engage battalion personnel (danger areas)?
  - Where are the natural firing positions that medical platoon personnel can use to defend against enemy attack?
- *Cover and concealment.*
  - What routes within the AO offer cover and concealment for placement of the BAS or a treatment team?
  - Do the natural firing positions in the AO offer cover and concealment for the platoon or enemy?

- *Weather.* The medical platoon leader can use these questions as he analyzes the impact of weather on the mission:

  - What are the light conditions (including percentage of night illumination) and visibility?

  - What are the times for beginning of morning nautical twilight (BMNT), sunrise, sunset, end of evening nautical twilight (EENT), moonrise, and moonset?

  - How has recent weather affected the AO?

  - How will fog, rain, dust, heat, snow, wind, or blowing sand affect the men and equipment during the mission?

NOTE

The effects of weather on smoke or NBC weapons/operations should also be considered.

(*d*) *Troop.*

- What is the supply status of ammunition, fuel, and other necessary items including Class VIII?

- What is the present physical condition of the soldiers, as well as of vehicles and equipment?

- What is the training status of the platoon?

- What is the state of morale?

- How much sleep have the men had?

- How much sleep will they be able to get before the operation begins?

- Does the platoon need any additional assets to support or accomplish its mission?

- What attachments are available to help the platoon accomplish its mission? What is the task organization in the WARNO or OPORD?

(*e*) *Time available.*

- What times were specified by the commander in the OPORD for such activities as movement, RECON, rehearsals, and LOGPAC operations?

- What priorities of work can the platoon accomplish (examples include security, maintenance, resupply, coordination, rehearsals, inspections, and sleep)?

NOTE

The medical platoon leader conducts reverse planning to ensure that all specified, implied, and essential tasks can be accomplished in the time available. He develops a reverse planning schedule (time line) beginning with actions on the objective and working backward through each step of the operation and preparation to the present time. This process also helps the platoon in making efficient use of planning and preparation time. Once the METT-TC analysis is complete, the medical platoon leader can then write the platoon mission statement based on the battalion CHS plan and answer the questions of WHO, WHAT, WHEN, WHERE, and WHY. This is a clear, concise statement of the purpose of the operation and the essential task(s) that will be crucial to its success. The essential tasks (the WHAT) should be stated in terms that relate to enemy forces, friendly forces, and/or the terrain (for example, "ESTABLISH BAS (-) ONE OR TWO TERRAIN FEATURES FROM THE MAIN BATTLE AREA"; "ESTABLISH THE BAS 1 TO 3 KILOMETERS FROM THE MAIN BATTLE AREA"; or "BE PREPARED FOR RAPID FORWARD DEPLOYMENT OF A TREATMENT TEAM IN SUPPORT OF COMPANY A"). The purpose (the WHY) explains how the platoon mission supports the commander's intent. The elements of WHO, WHERE, and WHEN add clarity to the mission statement.

NOTE

Simultaneous planning and preparation are key factors in effective time management during the troop-leading procedures. The next five steps (issue a WARNO; make a tentative plan; initiate movement; conduct RECON and coordination; and complete the plan) may occur simultaneously and/or in a different order. There may be multiple WARNOs.

b. *Issue a Warning Order.* The medical platoon leader alerts his platoon to the upcoming operation by issuing a WARNO that follows the five-paragraph OPORD format (see Table 3-1).

NOTE

The medical platoon leader will often do this from the battalion TOC during mission analysis.

Warning orders maximize subordinates' planning and preparation time by providing essential details of the impending operation and detailing major time line events that will support mission execution. The amount of detail included in a WARNO depends on the available time, the platoon's communications capability, and the information subordinates need to initiate proper planning and preparation. The WARNO may include the following information:

- Changes to task organization.
- Updated graphics (platoons equipped with intervehicular information systems or appliqué digital systems send new overlays).
- Enemy situation.
- Battalion mission.
- Commander's intent (if available).
- Combat health support mission.
- A tentative time line, to include the following:
  - Earliest time of movement.
  - Readiness condition and vehicle preparation schedule.
  - Reconnaissance.
  - Training/rehearsal schedule.

NOTE

Some individual and collective training may be initiated by the medical platoon leader before he issues the OPORD; this technique maximizes preparation time and allows the platoon to focus on tasks that will support the anticipated operations. For example, a medical platoon may train on treatment of different types of wounds or injuries that may be seen during the operations.

  - Time and location at the battalion OPORD will be issued, plus the platoon OPORD will also be briefed.
  - Service support instructions (if not included in the time line).

As critical information is received or updated, the medical platoon leader should issue subsequent or updated WARNOs to keep the platoon informed.

c. Make a Tentative Plan. The medical platoon leader begins developing his CHS plan when the battalion receives its first WARNO from the brigade. Based on the commander's intent, guidance, and the results of his mission analysis, the medical platoon leader develops a tentative plan that addresses all specified, implied, and essential tasks using the OPORD format. The tentative plan also covers RECON and coordination requirements between the platoon and adjacent and supporting units. The field medical assistant and the medical platoon SGT are excellent sources of ideas concerning the battalion CHS plan. The medical platoon leader can develop his COA and OPORD almost simultaneously with the battalion OPORD. Since the medical platoon is a battalion asset, most of their specified tasks will be developed during the battalion MDMP. The medical platoon leader can issue a WARNO from the battalion TOC during his participation in the MDMP. By the time the OPORD is given, he should have most of his platoon order prepared. The development of the platoon time line from backward planning should be his priority as soon as the battalion OPORD is prepared.

d. Initiate Movement. Many of the battalion-level operations require movement to forward assembly areas and to battle positions during the planning phase of an operation. This means that elements of the medical platoon will also move to these locations in support of the battalion units. Medical platoon elements move with the supported units according to the battalion plan. As often as possible, within the restraints of OPSEC, medical task organization should be done during hours of daylight, or as early as possible. In spite of multiple "own the night technologies," executing movements required by task organization is still more efficiently accomplished in daylight. Activities may include ensuring CHS for the company quartering party or beginning priorities of work.

e. Conduct Reconnaissance and Coordination. Effective RECON takes into account the factors of METT-TC and OAKOC from both friendly and enemy perspectives. As a minimum, the field medical assistant or the platoon SGT conducts a detailed map RECON to identify primary and alternate routes of MEDEVAC and preplanned sites for locating the BAS or a BAS (-). If time and security considerations permit and authorization is obtained from higher headquarters, an on-site ground RECON is the best way to survey the AO. In addition, the medical platoon leader should check with the S2 for an intelligence update. The medical platoon leader should take as many ambulance squad leaders as possible on his RECON. For offensive operations, the medical platoon leader should attempt to find a vantage point that will permit rapid accessibility to supported units while making use of natural terrain features to afford as much protection as possible for the BAS. Ground RECON for offensive operations usually is limited to checking routes to the start point (SP), the line of departure (LD), and the axis just beyond the LD. For defensive operations, the medical platoon leader should conduct a RECON of the unit position and the MEDEVAC routes to be used. Whenever tactically feasible, the platoon should make provisions to mark routes and locations for day and night operations during the RECON. See Chapter 4 for TTP on marking routes and positions. During the RECON (or during battalion-level rehearsals), the medical platoon leader or his representative should coordinate evacuation routes with supported companies.

f. Complete the Plan. The medical platoon leader refines the plan based on the results of the wargame, RECON, and coordination with the BSS' FSMC and supported battalion units. He then completes the plan using these results and any new information from his commander. He should keep the plan as simple as possible, at the same time ensuring that the CHS plan supports the commander's intent.

g. Issue the Order. Prior to the order, the platoon NCO should ensure that all subordinate elements have copies of the overlays correctly posted to the maps and a copy of any matrices from the battalion OPORD that covers key CHS actions. All platoon members receiving the OPORD verbally should be prepared to copy these instructions in the five-paragraph OPORD format. They should ask for explanations of any terminology or actions that they do not understand. If possible, the medical platoon leader issues the order from a vantage point overlooking the terrain on which the platoon will support the maneuver units. If not, he uses a terrain model, sand table, sketches, or his map to orient the platoon. He can also build a model of the AO using a briefing kit that contains such items as engineer tape, colored yarn, 3- by-5-inch index cards, and "micro" vehicle models. When time and security permit, the medical platoon leader issues the order to as many members of the platoon as possible. As a minimum, he assembles the treatment teams and ambulance squads, and combat medic section. He briefs the platoon using the five-paragraph OPORD format. He should then send the ambulance emergency care SGT to brief the trauma specialist assigned to each of the maneuver companies. To ensure complete understanding of the operation, the medical platoon leader and medical platoon personnel conduct confirmation briefings immediately after the OPORD is issued. The treatment team and ambulance team leaders brief the medical platoon leader to confirm their understanding of his intent, the specific tasks their team must perform, and the relationship between their tasks and those of other medical units/elements in the operation. If time permits, the medical platoon leader should lead the medical platoon in a walk-through using a sand table.

h. Supervise and Refine. Flexibility is the key to effective operations. The medical platoon leader must be able to refine his plan whenever new information becomes available. If he adjusts the plan, he must inform the platoon and supervise implementation of the changes. Once the operation has begun, the medical platoon leader must be able to direct his platoon in response to new situations and new orders. Platoon orders, back-briefs, rehearsals, and inspections are essential elements of the supervision process as the platoon prepares for the mission. The following paragraphs discuss these procedures in detail:

(1) *Team orders.* The medical platoon leader, the field medical assistant, and the platoon SGT make sure all ambulance team members have been briefed by squad leaders (heavy battalion and mechanized infantry medical platoons have track commanders [TC] for their M113 and M557 armored ambulance and treatment vehicles) and understand the platoon mission and concept of the operation.

(2) *Back-briefs.* The back-brief is, in effect, a reverse briefing process; those who receive an OPORD confirm their understanding of the order by repeating and explaining details of the operation for their leader or commander. In the medical platoon, the medical platoon leader should conduct back-briefs after the TC/team leaders have had a chance to review the OPORD but before the platoon rehearsal begins. The TC/team leaders brief the medical platoon leader on **how** their teams will accomplish the specific tasks assigned to them in the order.

NOTE

Although the back-brief is an effective means of clarifying the specifics of the plan, it does not require medical platoon personnel to practice or perform their assigned tasks. By itself, therefore, it is not an ideal rehearsal technique.

(3) *Rehearsals*. A rehearsal is a practice session conducted to prepare units for an upcoming operation or event. The medical platoon leader should never underestimate the value of rehearsals. Many units, in fact, consider rehearsals as a separate step (ninth) of troop-leading procedures. The medical platoon leader uses well-planned, efficiently run rehearsals to accomplish the following:

- Reinforce training and increase proficiency in critical tasks.
- Reveal weaknesses or problems in the plan.
- Synchronize the actions of subordinate elements.
- Confirm coordination requirements between the platoon and supporting medical units/elements.
- Improve each soldier's understanding of the concept of the operation, the direct fire plan, anticipated contingencies, and possible actions and reactions for various situations that may arise during the operation.

Effective rehearsals require personnel to perform required tasks, ideally under conditions that are as close as possible to those expected for the actual operation. Participants maneuver their actual vehicles or use vehicle models or simulations while interactively verbalizing their elements' actions. In a platoon-level rehearsal, the medical platoon leader will select the tasks to be practiced and will control execution of the rehearsal. The platoon can prepare for operations using reduced-force rehearsals and/or full-force rehearsals. The medical platoon leader conducts reduced-force rehearsals when time is limited or the tactical situation does not permit everyone to attend. Platoon members, who can take part, practice their actions on mock-ups, sand tables, or actual terrain (usually over a smaller area than in the actual operation). The full-force rehearsal is the most effective, but consumes the most time and resources. It involves every soldier who will participate in the operation. If possible, it should be conducted under the same conditions (such as weather, time of day, and terrain) that the platoon expects to encounter during actual operations. The medical platoon leader can choose among several techniques in conducting rehearsals, which should follow the crawl-walk-run training methodology to prepare the platoon for increasingly difficult conditions. Rehearsal techniques include the following:

- *Special rehearsal.* Individual and/or team tasks that will be critical to the success of the operation are rehearsed as necessary. The medical platoon leader may initiate special rehearsals when he issues the WARNO.
- *Map rehearsal.* This is usually conducted as part of a back-brief involving the TC or a complete ambulance team. The leader uses the map and overlay to guide participants as they back-brief their role in the operation. If necessary, he can use a sketch map.
- *Communications rehearsal.* This reduced-force or full-force rehearsal is conducted when the situation does not allow the platoon to gather at one location. Crewmen check their vehicles' communications systems and rehearse key elements of the platoon fire plan.

• *Key leader rehearsal.* Usually conducted as part of a larger force, this rehearsal involves leaders moving over the key terrain in wheeled vehicles while discussing the mission.

• *Sand table or terrain model.* The reduced-force or full-force technique employs a small-scale table or model that depicts graphic control measures and important terrain features for reference and orientation. Participants walk or move "micro" vehicles around the table or model to practice the actions of their own vehicles in relation to other members of the platoon.

• *Force on force.* This is used during a full-force rehearsal. The platoon may rehearse with sections or individual tanks going "force on force" against each other. Platoons should first rehearse with good visibility over open terrain. Rehearsals become increasingly realistic until they approximate those expected in the AO.

(4) *Inspections.* Inspections allow the medical platoon leader to check the platoon's operational readiness. The key goal is to ensure that soldiers and vehicles are fully prepared to execute the upcoming mission. Inspections also contribute to improved morale. It is essential that all leaders in the medical platoon know how to conduct precombat checks according to the platoon SOP. Procedures for a comprehensive inspection include the following:

• Perform before-operation maintenance checks and report or repair deficiencies.

• Upload vehicles according to platoon SOP. The standardization of load plans allows the medical platoon leader, the field medical assistant, and the platoon SGT to quickly check accountability of equipment. It also ensures standard locations of equipment in each vehicle; this can be an important advantage if the soldiers are forced to switch to a different vehicle during an operation.

• Review the supply status of rations, water, fuel, oil, ammunition, MES, first-aid kits, and batteries (for such items as flashlights, night vision devices, and NBC alarms). Direct resupply operations as necessary.

• Ensure vehicles are correctly camouflaged so they match the AO.

The medical platoon leader, the field medical assistant, and/or the platoon SGT should observe treatment and ambulance teams during preparation for CHS operations. They should conduct the inspection once the TC/squad or team leader report that their team(s) and vehicles are prepared. The precombat inspection must be a "hands on, show me that it works" event. This is the only way to ensure the platoon is properly prepared. If trauma specialists and/or ambulance teams cannot return to the platoon, then someone goes to inspect them.

3-5. Abbreviated Troop-Leading Procedures

When there is not enough time to conduct all eight troop-leading steps in detail, such as when a change of mission occurs after an operation is in progress, the medical platoon leader must understand how to trim the procedures to save time. Most steps of these abbreviated troop-leading procedures are done mentally, but the medical platoon leader skips none of the steps. Once the order is received, the medical platoon leader conducts a quick map RECON, analyzes the mission using the factors of METT-TC, and sends for the TC and team leaders. He makes sure they post the minimum required control measures on their maps and issues a FRAGO, covering the key elements of the enemy and friendly situations, the platoon mission, and the concept of the operation. The medical platoon leader and TC may also conduct a quick walk-through rehearsal of critical elements of the CHS plan using a hastily prepared terrain model or sand table. In some cases, there may not be enough time even for these shortened procedures. The platoon may have to move out and receive FRAGO by radio or at the next scheduled halt. It then becomes critical for the medical platoon leader to send FRAGO of his own to the TC and team leaders explaining the platoon's requirements and objectives in supporting the battalion maneuver plan. Digital and Global Positioning Systems (GPS) are valuable tools when the platoon is forced to use abbreviated troop-leading procedures and FRAGO. They allow the medical platoon leader to designate way points to assist in navigation and identifying evacuation routes. Other keys to success when abbreviated procedures are in effect include a well-trained platoon; clearly developed, thoroughly understood SOP; and an understanding by all members of the platoon of the current tactical situation (situational understanding). The medical platoon leader, the field medical assistant, and the platoon SGT must keep the platoon informed of the ever-changing enemy and friendly situations. They accomplish this by monitoring the battalion net and issuing frequent updates to the other platoon members using the radio and digital information systems. Whenever time is available, however, there is no substitute for effective, thorough troop-leading procedures. The odds of success increase still further when detailed planning and rehearsals are conducted prior to an operation, even if time is limited. Successful medical platoon leaders make the most of every available minute. Specified delegation of tasks will also speed this process up.

Section III. COMPUTERS, INFORMATION, DIGITIZATION, AND COMMUNICATIONS

3-6. Medical Standard Army Management Information System

The MC4 system when fielded will be a theater, automated CHS system, which links commanders, health care providers, and supporting elements, at all echelons, with integrated medical information. The MC4 system when developed and fielded will receive, store, process, transmit, and report medical C2, medical surveillance, casualty movement/tracking, medical treatment, medical situational understanding, and MEDLOG data across all levels of care. This will be achieved through the integration of a suite of medical information systems linked through the Army data telecommunications architecture. The MC4 system begins with the individual soldier and continues throughout the health care continuum. The best way to visualize the future MC4 system capability is as a piece of the Army digital computer network where all ten CHS functional areas have been digitized and CHS information is available to specified commands, supported units, and their personnel. See FM 8-10 for information on AMEDD functional areas. When fully developed, not only will the MC4 system provide Army commanders with CHS information, but will provide them with a seamless transition to the joint CHS environment. The MC4 system will consist of three basic components: software, hardware, and telecommunications systems.

a. Software capability.

(1) The joint TMIP will provide government off-the-shelf (GOTS)/commercial off-the-shelf (COTS) software and interoperability standards to support joint theater operations. The software provides an integrated medical information capability that will support all echelons of care in a theater of operations with links to the sustaining base. Medical capabilities provided by the software to support commanders in the theater will address medical C2 (including medical capability assessment, sustainability analysis, and MI); MEDLOG (including blood product management and medical maintenance management); casualty evacuation; and health care delivery.

(2) The MC4 system will support Army-unique requirements and any software needed to interface with Army information systems such as CSSCS, Global Combat Support System-Army (GCSS-A), FBCB2, Warrior Programs, and the Movement Tracking System. These systems will also be used in the medical platoons of the new IBCT. For additional IBCT medical platoon information, see Appendix D.

b. Hardware Systems. The hardware will consist of COTS automation equipment supporting the above software capabilities. Examples include, but are not limited to, computers, printers, networking devices, a digital patient record, and personal information carriers (PIC) that contain medical information.

c. Telecommunications Systems. The MC4 system will rely on current and proposed Army solutions for tactical, operational, and strategic telecommunications systems to transmit and receive digitized medical information throughout the theater and back to the sustaining base. There will be no separate AMEDD communication system. Telecommunications at brigade and below will be accomplished through the tactical internet; above brigade level, telecommunications will be accomplished through the Warfighter Information Network (WIN) architecture. The MC4 system will include hardware or software required to interface with current and emerging technologies supporting manual, wired, and wireless data transmission. At end-state, the MC4 system users will exchange data electronically via the WIN architecture. In the interim, commercial satellite and/or high frequency radios will be fielded to selected medical units (for example, Medical Detachment-Telemedicine [MDT], and so forth) receiving the MC4 system to support high bandwidth requirements until the WIN architecture is fully fielded. Personnel operating satellite assets are resourced in the MDT TOE and will be located with the MDT.

d. Patient Treatment Recording System. In the future under the MC4 system, medical information about each soldier of the maneuver battalion will be entered into a local database maintained at the supporting BAS or troop medical clinic. This information will include the soldier's immunization status, medical deployability status, and dental deployability status. Until a digital patient record and the PIC are fully functional and fielded, and in accordance with AR 40-66, a field medical record jacket (DD Form 2766) and its accompanying records will be maintained by the soldier's primary care provider. See Appendix B for definitive information on management of the individual health record in the field.

3-7. Information and Communications

a. The Medical Platoon Communications and Information Systems. Information and communications assets available to the BSS include those identified in Table 3-2. In the digitized medical platoon, each vehicle will have a FBCB2/position/navigation system (1 each). The FBCB2 system aids the medical platoon with maintaining real-time situational understanding of what is happening on the battlefield. The medical platoon exchanges information with forward deployed medical platoon elements, the FSMC, and the BSS.

b. Combat Health Support Functions on Force XXI Battle Command Brigade and Below System. The FBCB2 system is a hardware/software suite that digitizes C2 at brigade and below level. This system provides a seamless battle command capability for performance of missions throughout the operational continuum at the tactical level. The FBCB2 system is the implementation of information technology to provide increased battlefield operational capabilities. The system is positioned on specified platforms and will perform combat, CS and CSS functions for the planning and execution of operations. This system gives the medical platoon a CRP of the current CHS situation at BASs, AXPs, and the FSMC. For the first time, the medical organizations and elements are digitally linked to the platforms and organizations they support. The current CSS functionality on FBCB2 system gives the combatant a common relevant picture of the current CSS situation at his level of command and at subordinate levels. It also provides the personnel and logistics leaders situational understanding of CSS operations throughout their battle space. It provides an enhanced capability to synchronize support to customer units. Combat service support functionally on FBCB2 includes the following:

- Logistical Situational Report (LOGSITREP).
- Personnel Situational Report.
- Medical Situational Report (MEDSITREP).
- Situational understanding.
- Logistics call for support.
- Logistics task order.

Currently, the FBCB2 system also permits information to be entered using free text such as comments and other pertinent CSS information. This common battle space picture will enable CHS providers to maintain the operational tempo set by the maneuver commander. There are three medical screens incorporated into the CSS FBCB2 function. They are the medical functionality in the LOGSITREP, the MEDSITREP, and the MEDEVAC request. It is important that units use standard message and report formats to eliminate confusion. As the system is further developed and additional CHS screens are added, there will be less space for using free text. Figure 3-3 is the medical screen as seen on the CSS function of FBCB2. Descriptions of each screen are provided below.

(1) *Medical functionality in the logistical situational report.* This message provides visibility of selected Class VIII items at the BAS and FSMC stock levels, date and time group of the most recent report, and location of medical units. Recipients of the report are the FSC, the FSB support operations (HSSO), the BSS, and the DSS. This report is entered into the CSSCS by the BSS.

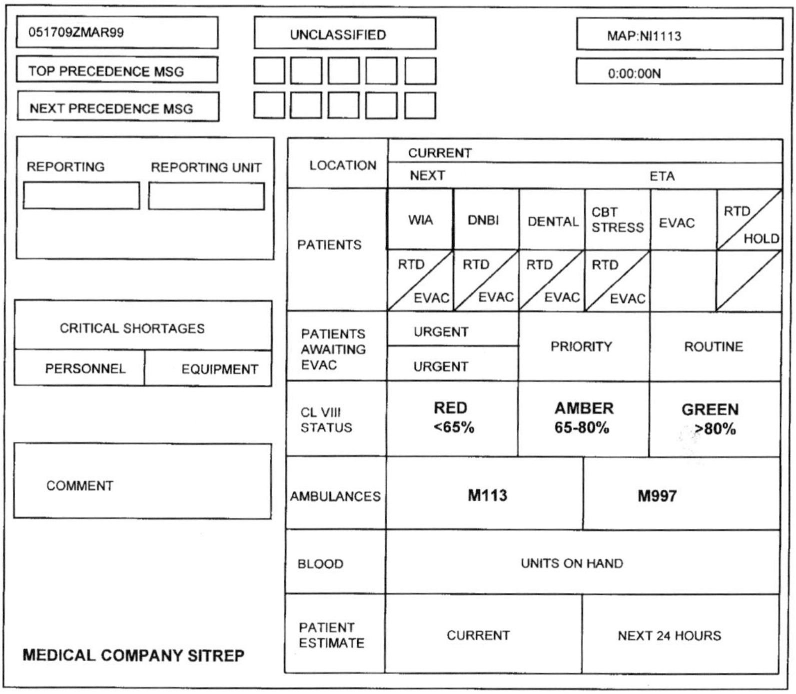

Figure 3-3. Sample of medical screen incorporated into the combat service support Force XXI Battle Command Brigade and Below function.

(2) *Medical Situational Report.* The FSMC and BAS prepare and submit this report. The recipients of the report are maneuver commanders and their S1, the FSC, the FSB support operations (HSSO), the BSS, and the DSS. The BSS and FSMC receive roll up from the BASs. The DSS receives a roll up of the FSMC reports. Adjacent units can receive information copies of the MEDSITREP. This message reports the following information:

(*a*) Current location and proposed next location with estimated time of arrival.

(*b*) Number of patients seen and classified as wounded in action, DNBI, dental, and combat stress. The field will also show the number of patients evacuated and the number RTD.

(*c*) Patient(s) awaiting MEDEVAC.

(*d*) The Class VIII status of the element/unit, the number of ambulances that are mission capable, and the number of units of blood and type on hand, if required.

(*e*) There will be a free text field for critical Class VIII or other supply shortages and commander's comments.

(3) *Medical evacuation request.* This request is currently embedded into FBCB2 and is a digitized standard 9-line MEDEVAC request. The current messaging is from the requestor to the medical platoon leader (with an information copy to the maneuver battalion commander). The medical platoon leader either responds or forwards the request to the FSMC commander who dispatches the appropriate MEDEVAC asset. Information copies of all MEDEVAC requests are sent to the BSS so they can maintain real-time situational understanding on the volume of requests. The FSMC commander sends an information copy to the BSS with after-action information that includes destination of evacuated patient(s).

3-8. Radio Nets

a. *Battalion Communications.* Battalion communications are sent over a variety of radio nets. Primary battalion communications nets are—

(1) *Command net.* A secure command net is used for C2 of the TF. All organic and attached units, including the FSO, forward air controller, and leaders of supporting elements, enter the battalion command net. Primarily, during the execution of the mission, only commanders transmit; all others monitor and transmit only essential information. The command operations net (see Figure 3-4) is controlled by the battalion main CP.

(2) *Operation and intelligence net.* The operations and intelligence (O&I) net is a secure net established to provide a mechanism for the battalion TF to accept routine items of information concerning O&I reporting without cluttering or interfering with the battalion command net.

(3) *Administrative/logistics net.* The administrative/logistics net is a tactical net, controlled by the combat train command post (CTCP), used to communicate the administrative and logistical requirements of the TF. All organic and attached units normally operate in this net.

(4) *Special radio nets.*

(*a*) The scout platoon net or a designated frequency may function as a surveillance net when required. The S2 and elements assigned surveillance missions operate on this net. Other elements enter or leave the net to pass information as required.

(*b*) The FSE and company fire support teams operate in the supporting FA command fire direction net and a designated fire direction net to coordinate FA fires for the battalion. The TACP operates in USAF tactical air-request and air-ground nets to control air strikes.

(*c*) Supporting air defense units monitor the early warning net. In the absence of collocated air defense support, the main CP will also monitor the division early warning net.

(*d*) Attached or OPCON support assets may operate in their parent unit nets, but they must also monitor the command net at all times.

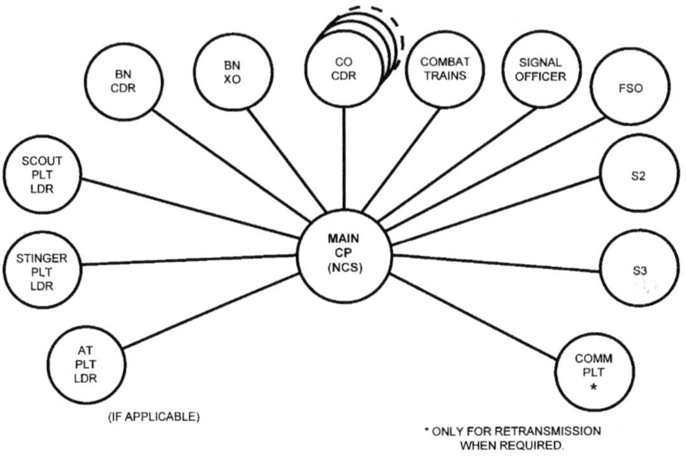

Figure 3-4. Example of a battalion command operations net.

b. Administrative/Logistics Radio Net. The administrative/logistics radio net (see Figure 3-5) is used for most CSS traffic. However, at battalion level, CSS communications can be via any combination of FM radio, mobile subscriber equipment (MSE), courier, computer, or wire. Lengthy reports should be sent by messenger, wire, computer, or tactical facsimile (FAX).

(1) The CTCP is the NCS for the administrative/logistics net. The S4, S1, HHC commander, battalion maintenance team (less light), support platoon leader, medical platoon leader, company XO or 1SG, and others (as required) operate in the battalion administrative/logistics net. The CTCP also operates in the brigade administrative/logistics net and in the battalion command net.

(2) The main CP and CTCP should be positioned, when wire is available and circumstances permit, so wire can be used as the main means of communication between them. Wire allows a constant flow of information between the CP. It also enhances the ability of the CTCP to stay abreast of the tactical situation and thus to provide better support. Wire communications produce no electronic signature and, therefore, are more secure than radio. When MSE is fielded, wire is needed only as a backup means of communications.

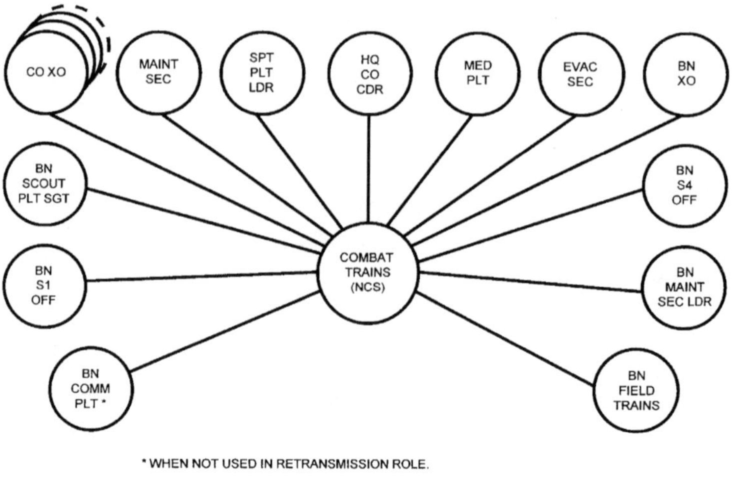

Figure 3-5. Example of an administrative/logistics net.

c. Medical Platoon Internal Operations Net. The platoon has access to the maneuver battalion HHC wire communications network for communications with all major elements of the battalion. Wireless communications for this section consists of a tactical FM radio mounted in the platoon headquarters vehicle. The medical platoon employs an FM radio network for CHS operations. The platoon headquarters section serves as the NCS for the platoon (see Figure 3-6). Table 3-2 lists the information and communications assets available to the Force XXI medical platoon.

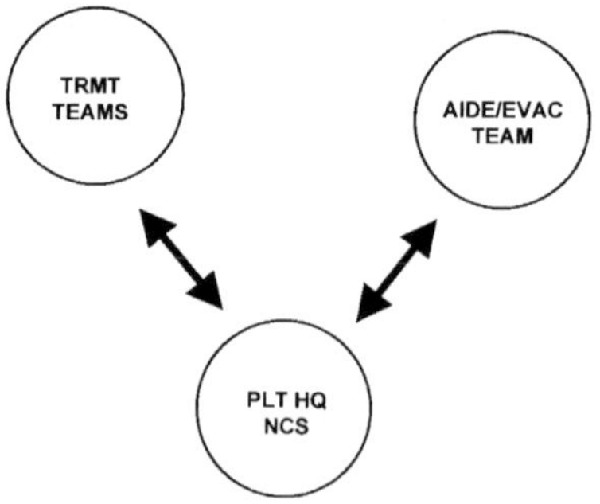

Figure 3-6. Example of the medical platoon internal operations net.

Table 3-2. Information and Communications Assets Available to the Force XXI Battalion Medical Platoon

RADIO SETS	
AN/VRC-89F	PLATOON HEADQUARTERS VEHICLE AND TREATMENT TEAM ALPHA (SURGEON) VEHICLE (1 EACH)
AN/VRC-88F	TREATMENT TEAM BRAVO
AN/VRC-90F	PLATOON AMBULANCE (1 EACH)
ROUTERS	**COMPUTER SYSTEMS**
TACTICAL LAN LAN ROUTER	MC4 LAPTOPS FOR EACH TREATMENT AND AMBULANCE VEHICLE
	FBCB2, 1 IN EACH VEHICLE ASSIGNED TO THE MEDICAL PLATOON
	OTHER SYSTEMS
FBCB2	MEDICAL PLATOON VEHICLE (1 EACH)
GPS	MEDICAL PLATOON VEHICLE (1 EACH)
EPLRS	MEDICAL PLATOON VEHICLE (1 EACH)
BCIS	MEDICAL PLATOON VEHICLE (1 EACH)
DVE	MEDICAL PLATOON VEHICLE (1 EACH)

d. Brigade Medical Operations Net. The brigade medical operations net under Army of Excellence units is the FSMC's command net. For Force XXI brigades and brigades with surgeon's sections assigned to the brigade, the brigade headquarters will maintain communications with medical elements supporting the brigade through its FM medical net. Single-channel ground and airborne radio system (SINCGARS) components provide the BSS with an AN/VRC-89 series radio set (FM) which has a receiver/transmitter (RT) capable of using two FM nets for reception and transmission. This permits the BSS to communicate with CHS elements via the administrative/logistic net (FM). The AN/VRC-89 series has two RTs (and one power amplifier). Two RTs allow the BSS to participate in two FM nets. These nets include the brigade administrative/logistics net and one each of the three medical platoons operations nets. The BSS also communicates using amplitude modulated (AM)-improved high frequency radios (IHFR) with its AN/GRC-213 or AN/GRC-193A radio. Another technique is to use the FSMC command net for brigadewide medical communications while using the administrative/logistics net for other CSS integration (see Figure 3-7). Situational understanding is monitored using the FBCB2 system and by face-to-face contact with other brigade staff members in the brigade TOC.

e. Mobile Subscriber Equipment. Mobile subscriber equipment will allow the BSS to communicate throughout the battlefield in either a mobile or static situation. As the Army continues to digitize the battlefield and modernize the force, the use of automation continues to develop. The MSE packet switching network gives units the ability to connect to division and corps LAN or wide area networks (WAN). A WAN is similar to the LAN but covers a larger distance. This allows units/CP to connect computer systems such as the CSSCS, maneuver control system, and FBCB2 system to an ethernet cable (coaxial) and send and receive information in an extremely efficient manner. Because of the limitations of a network constructed with coaxial cable, a WAN uses a combination of the MSE packet switch network and radio networks to distribute the data where necessary through the system. Packet switching does not use or take up existing telephone lines. Instead, telephone lines are freed up even more because information is being sent over a network on computers and related equipment. Using the common hardware/software facilitates the interface and exchange of information between the BSS and the medical platoons operating BAS, the FSMC, the DSS, the corps, and the division medical elements. See FM 63-2-2 for information concerning automated data processing continuity of the operations plan.

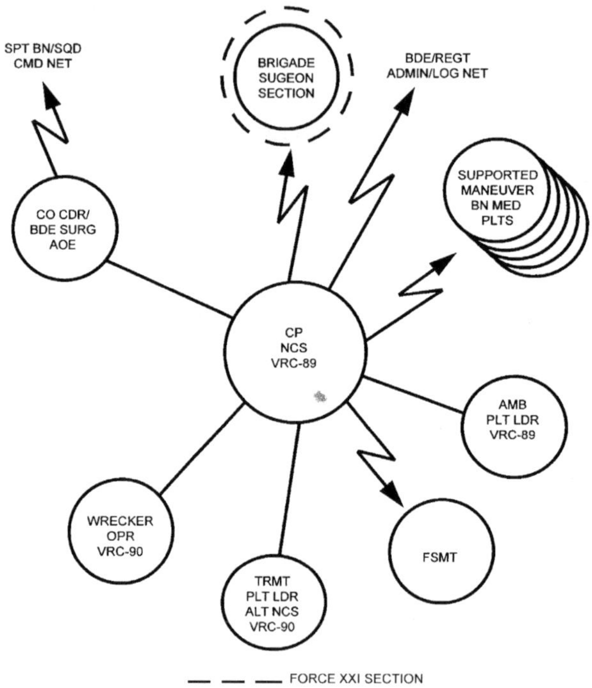

Figure 3-7. Medical company/troop command FM net/brigade medical operations net.

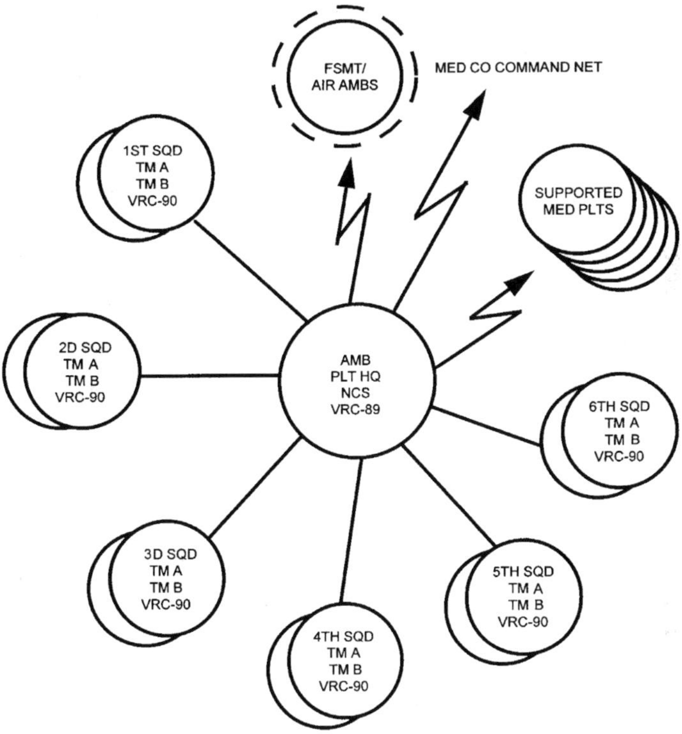

Figure 3-8. Dedicated medical evacuation FM net.

CHAPTER 4

COMBAT HEALTH SUPPORT OPERATIONS

4-1. Combat Health Support for Reconnaissance Operations

Reconnaissance precedes all military operations and verifies or refutes analyzed information in IPB products. The RECON information may require that the initial plans or COA be modified or discarded. The R&S systems are used to see the enemy, terrain, and other aspects of the battle space that will affect operations. As advanced warfighting systems are fielded, the R&S capabilities at all levels will be significantly enhanced but will never completely replace human collection requirements. The scout platoon is the primary R&S asset for the battalion/TF. This unit is specifically designed and equipped to conduct R&S operations and can do some limited intelligence analysis. It is the only element in the battalion/TF specifically trained in reconnaissance and can work up to 30 kilometers across the forward line of own troops (FLOT) forward of the battalion/TF. The key to synchronizing this asset is predicated on each member of the platoon understanding the battalion/TF commander's operational intent and concept. Members should understand their roles in obtaining the major decision point criteria. Detailed plans and orders are key to effective employment of the scout platoon. The continuous nature of the R&S process requires the commander to manage the employment of the scout platoon for R&S requirements tied to critical decision points. Reconnaissance missions require a detail list of exactly what is being tasked and a clear priority for collection. If RECON assets could find and report only one piecė of information, what does the commander want it to be? For example, the answer to this questions is reflected in the instructions to the scouts. "Execute zone RECON forward of the battalion/TF, beginning at 0345 hours. Collect and report the following in priority. Phase I: Number 1, confirm or deny enemy antitank minefield at grid FT 456689; Number 2, determine enemy unit and type and disposition of hill 413, grid FT535731." This allows the scout platoon leader to collect vital information first and focus his efforts, enhancing the battalion/TF's planning process. Reconnaissance assets should not be launched without determining the initial RECON plan. The medical platoon leader will be involved with developing the CHS portion of the RECON plan. The RECON plan can be issued as a FRAGO and should include—

- Composition/task organization for RECON.
- Key facts (priority intelligence requirements) to be gathered by the RECON party and expected results.
- Movement routes/formations to the RECON location.
- Actions on reaching the location.
- Special instructions to members of the RECON party. Collection task should be specified.
- Any special equipment required (chemical detection kits, expedient or directional antennas, and so on).
- Contingency plans.
- Requirements for continued surveillance after the RECON.

- Indirect fire support for RECON.
- Security arrangements.
- Communications arrangements (nets, retransmission sites, and so on).
- Withdrawal, linkup, and dissemination plans.

Combat service support considerations for RECON include—

- Developing SOPs for load plans of outposts.
- Establishing CSS procedures for both mounted and dismounted RECON missions.
- Establishing CSS procedures for air versus ground insertion of RECON assets.
- Developing procedures for both aerial and ground sustainment.
- Developing resupply techniques to include—
  - Using multiple/false landing zones away from outposts as cache drop-off points.
  - Predetermining the locations and times for resupply of Classes I, III, IV, V, VII, and XI.
  - Establishing locations for caches on successive missions/insertions.

Combat health support considerations include—

- Determining the CHS requirements.
- Deploying trauma specialists and medical assets in DS of RECON operations, as required and appropriate.
- Planning for casualty evacuation/extraction operations.
- Developing a CHS SOP for supporting RECON elements deployed deep into enemy territory.
- Selecting evacuation sites (remembering that all cache sites are potential casualty evacuation sites).
- Developing TTPs for cross-FLOT casualty extraction.
- Establishing the time for pick up and the pickup point for aerial extraction of casualties. (The last known/reported location is normally the aerial pickup point and the best time is 30 minutes prior to BMNT or 30 minutes after EENT.)

- Developing a detailed plan for ground extraction, to include link up to quick reactionary force (QRF) and escort to casualty exchange point.

- Conducting rehearsals for day and night extractions.

- Requesting escort if a QRF is not established.

During RECON missions if a member or members of the platoon become casualties, initial care will be self-aid, CLS advanced first aid, or EMT from a trauma specialist. Trauma specialists may be deployed as a rider in one of the scout vehicle or an ambulance team may be in support. If casualties/patients require extraction/medical evacuation, it becomes a combat mission. Since this is a combat mission that may require offensive action, medical personnel are placed under the control of a QRF leader. Provided in Figure 4-1 is a technique for conducting cross-FLOT ground extraction and lists the overall considerations. Figure 4-2 is a technique for conducting cross-FLOT air extractions and lists the overall considerations. Since the RECON element is usually small, enemy contact could produce a significant number of casualties requiring extraction of the entire team. Either ground or air could be used to execute the extraction of the RECON team. As stated above, this is a combined arms operation. Medical personnel will participate in the planning, preparation, and execution of these missions. If medical evacuation vehicles are not sent with the extraction force, then one vehicle should be designated for casualties and augmented with a trauma specialist. For additional information on reconnaissance operations, see FMs 7-20, 7-30, 7-92, 17-98, and 71-3.

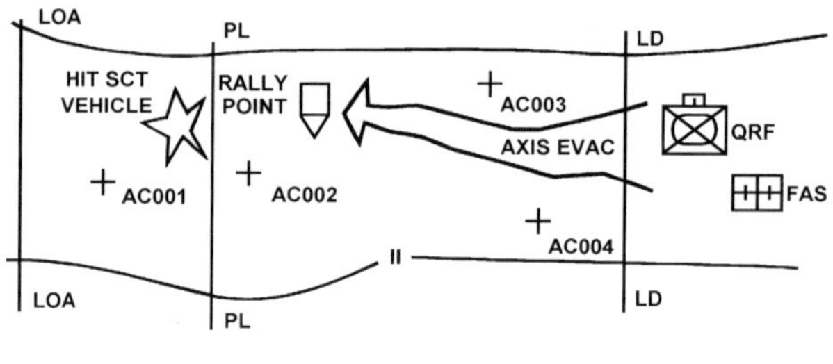

OVERALL CONSIDERATIONS

- EXTRACTION IS A COMBAT MISSION
- PLAN/REHEARSE QRF
- PLAN COMBAT MULTIPLIERS
- PLAN ARTILLERY SUPPORT THROUGHOUT ZONE/SECTOR
- PLAN FALSE/DISTRACTING ARTILLERY
- CLOSE AIR SUPPORT
- GRAPHICS ISSUED TO SUPPORT MISSION
- ALERT ALL RECON ASSETS
- USE RECON ASSETS IN SECTOR/ZONE TO VECTOR QRF TO RALLY POINT
- POSITION AID STATION NEAR QRF STAGING AREA
- PLAN MUST COVER MOVEMENT IN AND OUT OF SECTOR/ZONE
- LIMITED ATTACKS TO DISTRACT ENEMY/CLEAR ROUTES, ETC.

Figure 4-1. A technique for cross-FLOT ground extraction.

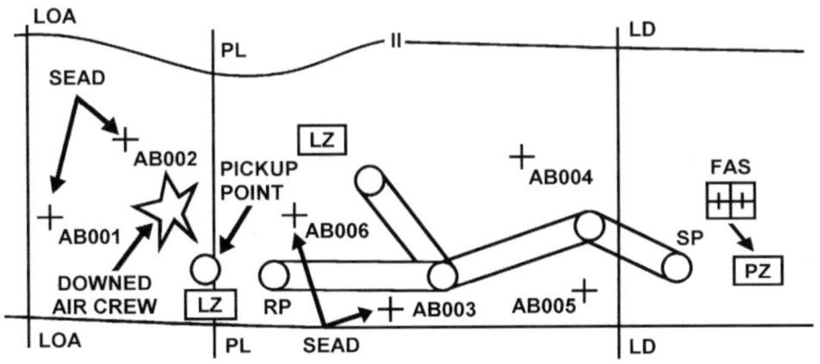

OVERALL CONSIDERATIONS

- EXTRACTION IS A COMBAT MISSION
- PLAN COMBAT MULTIPLIERS
- LETHAL SEAD—INGRESS/EGRESS
- NONLETHAL SEAD—TLQ-17/COMPASS CALL
- FALSE LZ/INSERTIONS
- FALSE ARTILLERY/SEAD
- CLOSE AIR SUPPORT
- ALERT ALL RECON ASSETS
- ISSUE INTEL UPDATE AT PZ
- USE RALLY POINTS/DOWNED CREW PICKUP POINTS
- POSITION AID STATION NEAR LZ

Figure 4-2. A technique for cross-FLOT air extraction.

4-2. Combat Health Support for the Offense and the Defense

a. Combat Health Support for the Offense.

(1) The offense is the decisive form of war, the commander's only means of attaining a positive goal, or of completely destroying an enemy force (FM 100-5). Rapid movement, deep penetrations, aggressive action, and the ability to sustain momentum regardless of counterfires and countermeasures characterize the offense.

(2) When developing the CHS plan to support the offense, the CHS planner must consider many factors (FM 8-55). The forms of maneuver, as well as the threat's capabilities, influence the character of the patient workload and its time and space distribution. The analysis of this workload determines the allocation of CHS resources and the location or relocation of MTF.

(3) Combat health support for offensive operations must be responsive to several essential characteristics. As operations achieve success, the areas of casualty density move away from the supporting MTF. This causes the routes of MEDEVAC to lengthen. Heaviest patient workloads occur during disruption of the threat's main defenses, at terrain or tactical barriers, during the assault on final objectives, and during threat counterattacks. The accurate prediction of these workload points by the CHS planner is essential if MEDEVAC operations are to be successful.

(4) As advancing combat formations extend control of the battle area, supporting medical elements have the opportunity to clear the battlefield. This facilitates the acquisition of the battle wounded and reduces the vital time elapsed between wounding and treatment. There are two basic problems confronting the supporting medical units and MEDEVAC elements. First, contact with the supported units must be maintained. Responsibility for the contact follows the normal CHS pattern—higher echelon evacuates from lower echelon. Contact is maintained by forward deployed air and ground evacuation resources. Secondly, the mobility of the MTF supporting the combat formations must be maintained. The requirement for prompt MEDEVAC of patients from forward MTF requires available ambulances to be echeloned well forward from the outset. Air and ground ambulance support beyond the capabilities of the FSMC is requested from the supporting corps MEDEVAC battalion.

(5) In traditional combat operations, the major casualty AO is normally the zone of the main attack. As the main attack accomplishes the primary task of the combat force, it receives first priority in the allocation of combat power. The allocation of combat forces dictates roughly the areas that are likely to have the greatest casualty density.

(*a*) In the division, CHS (Echelon I) for the brigade is provided by maneuver battalion medical platoons. The medical platoon operates the BAS or squadron aid station, places trauma specialists in DS of the maneuver companies and RECON troop, provides patient evacuation from forward areas, and deploys treatment teams in DS of battalion or squadron elements for up to 48 hours without resupply. The FSMC (Echelon II) located in the BSA provides MEDEVAC support from the BAS/squadron aid station to the BSA and reinforces treatment capabilities at BAS/squadron aid stations for limited periods of time. When combat operations commence, the medical platoon normally locates its BAS as far forward as combat operations permit. The BAS treatment squad can split into two treatment teams and operate as two separate aid stations (BAS [-]), normally not to exceed 24 hours. Treatment Team A operates the main aid station (MAS) and Treatment Team B operates the forward aid station (FAS). In continuous operations, when operating for longer periods, personnel efficiency and unit capability will tend to deteriorate. Each team employs treatment vehicle(s) with two MES—one trauma set and one general sick call set. The medical platoon will depend on CLS, trauma specialists, company health care SGT, and unit 1SG to assist with clearing the battlefield and getting injured or wounded soldiers to the CCP. From the forward areas and the CCP, armored ambulances will provide medical evacuation to the FAS and MAS. Based on casualty estimates, additional ambulances may be forward positioned at CCP. The CCP, Treatment Team A, and Treatment Team B must select sites that provide cover and concealment and afford some protection to their patients. For additional information on force protection, see Appendix E. Treatment teams must maintain as much mobility as possible while providing stabilization care and rapid medical evacuation. Treatment and evacuation elements must ensure that adequate stocks of Class VIII items are on-hand and, if necessary, request Class VIII resupply from the FSMC.

(*b*) The Force XXI medical platoon may deploy a treatment team forward in support of brigade RECON troops based on mission requirements. As the battle moves from the original area of contact, coordinated movements of treatment teams allow for continuous CHS. Once patients are received, a treatment team from the BAS or squadron aid station will care for and treat these patients until their MEDEVAC or appropriate disposition. The remaining treatment teams of the maneuver BAS and squadron aid stations move with the battle and provide CHS to the maneuver and RECON elements according to the brigade order.

(*c*) Careful coordination must be used when deciding when and which treatment team will receive patients. The plan must take into account maintaining the mobility of the medical platoon so that at least one treatment team will be able to follow the force as it maneuvers. This will keep evacuation routes and times to a minimum. Maneuver BAS/treatment teams provide area medical support to units without organic medical support operating in the maneuver battalion's AO. Brigade RECON will plan for area medical support for cross-FLOT operations. After MEDEVAC or appropriate disposition of their patients, the treatment team prepares for its next move. This echeloned displacement (leapfrog) technique provides for maximum utilization of medical platoon treatment teams and permits continuous uninterrupted CHS to maneuver battalions on the move. Each of the above actions must be coordinated with the CTCP and the FSMC.

(*d*) The BAS is under the tactical control of the battalion S4 and is normally deployed in the vicinity of combat trains. To reduce ambulance turnaround time in providing ATM to patients within 30 minutes of wounding, the BAS may split and place its treatment teams as close to maneuvering companies as tactically feasible. The battalion S4 closely coordinates locations for forward positioning CSS elements (including medical treatment elements) with the battalion S3. This is to ensure that the location of these elements is known by commanders of maneuvering and CS forces. Coordination ensures that CSS elements are not placed in the way of friendly maneuvering forces; in line of incoming or outgoing fires; or in areas subject to be overrun by rapidly advancing enemy forces. Treatment teams situated close to (within 1,000 meters of) maneuvering companies in contact must be prepared to withdraw to preplanned, alternate positions on short notice. They must also be aware of enemy capabilities. Some adversaries will possess weapons capable of direct fire at 4,000 meters.

(*e*) Trauma specialists are allocated to infantry and mechanized infantry companies on the basis of one trauma specialist per platoon and, under Force XXI, a health care SGT is allocated for each company. In armored units, the allocation is one health care SGT and one ambulance team per company. The location of the trauma specialist is of extreme importance for early acquisition and medical treatment of casualties. The mechanized infantry platoon trauma specialist normally locates with, or near, the element leader. When the platoon is moving on foot in the platoon column formation, he positions himself near the element leader trailing the base squad forward of the second team. This formation is the platoon's primary movement formation. When the platoon is mounted, the trauma specialist will usually ride in the same vehicle as the platoon SGT. The company senior trauma specialist collocates with the 1SG. When the company is engaged, he remains with the 1SG and provides medical advice as necessary. When a casualty occurs, first aid will be rendered by self-aid/buddy aid or by the CLS. The platoon/company trauma specialist will then go to the casualty's location or the casualty will be brought to the trauma specialist. The trauma specialist makes his assessment; administers initial medical care; initiates a DD Form 1380 (FMC); then requests evacuation or returns the individual to duty. The patient is evacuated from the point of injury or is transported on a nonmedical vehicle (coordinated by the 1SG) to a collection point. A vehicle from the medical platoon evacuation section (usually pre-positioned forward) picks up the patient and transports him to the BAS. As the tactical situation allows, the trauma specialist will provide medical treatment and prepare patients for evacuation. The ambulance team supporting the company works in coordination with the trauma specialists supporting the platoons. When a casualty occurs in a tank or an armored fighting vehicle, the evacuation team will move as close to the vehicle as possible, making full use of cover, concealment, and defilade. Assisted, if possible, by the vehicle's crew, they will extract the casualty from the vehicle and administer EMT. They move the patient to the treatment team or to a CCP to await further

MEDEVAC. The company health care SGT normally remains with the company CP, but may be used anywhere in the company, assisting the ambulance teams in some situations. The 1SG oversees the evacuation of casualties back to a CCP and may employ nonstandard medical platforms to accomplish his mission. The company health care SGT will position himself at the CCP and provide medical treatment, to include prioritizing patients for MEDEVAC to the BAS.

(6) When maneuvering companies anticipate large numbers of casualties, augmentation of the medical platoon with one or more treatment teams from the FSMC should be made. Augmenting treatment teams are under the tactical control of the battalion S4, but are under the technical control of the battalion surgeon. Medical treatment facilities should not be placed near targets of opportunity such as ammunition, POL distribution points, or other targets that may be considered lucrative by the opposing force. Considerations for the location of the BAS should include—

- Tactical situation/commander's plan.
- Expected areas of high casualty density.
- Security.
- Protection afforded by defilade.
- Convergence of lines of drift.
- Evacuation time and distance.
- Accessible evacuation routes.
- Avoidance of likely target areas such as bridges, fording locations, road junctions, and firing positions.
- Good hard stand on solid ground with good drainage.
- Near an open area suitable for helicopter landing.
- Available communication means.

(7) Types of operations in the offense include—

(*a*) *Movement to contact*. Medical evacuation support in movement to contact is keyed to the tactical plan. Prior deployment of FSMC ground ambulances with the maneuver battalion's organic medical platoons permits uninterrupted and effective MEDEVAC support from the BAS to the FSMC located in the BSA. Movement to contact operations is executed when there is little or no threat information. The FSMC and treatment teams from maneuver BAS in support of these operations must maintain their flexibility and be prepared to adjust CHS once contact is established.

(*b*) *Exploitation and pursuit*. Medical evacuation support of exploitation and pursuit operations resemble those discussed for the envelopment (paragraph 4-3*a*[2]). Since exploitation and pursuit operations can rarely be planned in detail, evacuation operations must adhere to TSOP and innovative C2. These actions are often characterized by—

- Fewer casualties
- Decentralized operations.
- Unsecured ground evacuation routes.
- Exceptionally long distances for evacuation.
- Increased reliance on convoys and air ambulances.
- More difficult communications.

(*c*) *Deliberate attack*. The deliberate attack is based on a more detailed knowledge of the threat disposition and likely actions. The brigade's actions in contact will be more predictable than the fluid situation found in the movement to contact, or exploitation or pursuit. Specific terrain and routes/ avenues of approach can be selected. Units can conduct at least a map RECON of their planned locations. While there may be CHS requirements during the approach, the assault on the objective will produce the greatest number of casualties. Some of the CHS considerations for the deliberate attack include—

- Higher percentage of casualties.
- Casualties will be more concentrated in time and space.
- Once the objective is secured, treatment teams can move to the objective instead of evacuating patients from the objective to the treatment teams.
- Use of air ambulance to overcome some obstacles may be required.
- Higher likelihood of wounded EPW.

(8) At the BAS, patients requiring further evacuation to the rear are stabilized for movement. Patient holding and food service are not available at the BAS; therefore, only procedures necessary to preserve life or limb, or enable a patient to be moved safely are performed at the BAS. Evacuation from the BAS or point of injury is performed by the FSMC's ambulance platoon and by the FSMT. Constant effort should be made to prevent unnecessary evacuation; patients with minor wounds or illnesses are treated and RTD as soon as possible. Patients with dental emergencies are evacuated to the FSMC for treatment. Ammunition, hand grenades, and light antitank weapons are not evacuated from the forward areas with a patient. Ambulatory patients may retain their rifle or pistol. Other weapons, such as machine guns and those items identified above, are collected and given to the soldier's 1SG or turned in to the battalion S4 for appropriate disposition according to command SOP/policy. All excess equipment collected

at the BAS is turned in to the battalion S4 or as directed by command SOP. Patients will always retain their protective mask. Other functions of the BAS include—

- Receiving and recording patients in the daily disposition log.

- Preparing FMC as required (see FM 8-10-6 for definitive information on initiating and completing a DD Form 1380).

- Verifying information contained on the FMC of all patients evacuated to the BAS.

- Requesting and monitoring MEDEVAC of patients.

- Monitoring personnel, when necessary, for NBC contamination prior to medical treatment.

- Notifying the S1 of all patients processed through the BAS, giving identification and disposition of patients.

- Supervising patient decontamination and treating NBC patients (see FMs 8-9, 4-02.283, 8-284, and 8-285).

NOTE

Patient decontamination is performed by a pretrained decontamination team. This team is composed of eight nonmedical personnel from supported units that work under the supervision of medical personnel. Patient decontamination teams perform best when they are trained and are permitted to exercise their skills with the supporting BAS. For definitive information on patient decontamination procedures, see FMs 8-10-1, 8-10-7, 4-02.283, 8-284, and 8-285.

b. Combat Health Support for the Defense. There are three forms of the defense: area defense, mobile defense, and retrograde. The area defense concentrates on denying threat access to designated terrain for a specific period of time, rather than on the outright destruction of the threat. The mobile defense focuses on denying the threat force by allowing him to advance to a point where he is exposed to a decisive counterattack by the striking force. The primary defeat mechanism, the counterattack, is supplemented by the fires of the fixing force. The third form of defense is the retrograde. The retrograde is an organized movement to the rear and away from the threat. The threat could force these operations or a commander can execute them voluntarily. Delay, withdrawal, and retirement are the three forms of retrograde operation.

(1) Combat health support provided for the defense may reflects lower casualty rates, but threat actions and the maneuver of combat forces complicate forward area patient acquisition.

Medical personnel are permitted much less time to reach the patient, complete vital EMT, and remove him from the battle site. Increased casualties among exposed medical personnel further reduce the medical treatment and evacuation capabilities. Heaviest patient workloads, including those produced by threat artillery and NBC weapons, can be expected during the preparation or initial phase of the threat attack and in the counterattack phase. The threat attack can disrupt ground and air routes and delay evacuation of patients to and from treatment elements. The depth and dispersion of the defense create significant time and distance problems for evacuation assets. Combat elements could be forced to withdraw while carrying their remaining patients to the rear. The threat exercises the initiative early in the operation, which could preclude accurate prediction of initial areas of casualty density. This makes the effective integration of air assets into the MEDEVAC plan essential. The use of air ambulances must be coordinated with the FSMT, normally positioned in the BSA under Force XXI and/or with either the aviation brigade or the BSA in AOE divisions. See FM 8-10-26 for additional information on the air ambulance company and FSMT.

(2) The CHS requirements for retrogrades can vary widely depending upon the tactical plan, the threat reaction, and the METT-TC factors. Firm rules that apply equally to all types of retrograde operations are not feasible, but considerations include—

- Requirement for maximum security and secrecy in movement.

- Influence of refugee movement conducted in friendly territory, which could impede MEDEVAC missions.

- Integration of evacuation routes and obstacle plans.

- Difficulties in controlling and coordinating movements of the force that could produce lucrative targets for the threat.

- Movements at night or during periods of limited visibility.

- Time and means available to remove patients from the battlefield.

In stable situations and in the advance, time is important only as it affects the physical well-being of the wounded. In retrograde operations, time is more important. As available time decreases, CHS managers at all echelons closely evaluate the capability to collect, treat, and evacuate all patients.

- Medical evacuation routes required for the movement of troops and materiel. This causes patient evacuation in retrograde movements to be more difficult than in any other type of operation. The threat could disrupt C3. Successful MEDEVAC requires including ambulances on the priority list for movement; providing for the transportation of the slightly wounded in cargo vehicles; and providing guidance to subordinate commanders defining their responsibilities in collecting and evacuating patients. Special emphasis must be placed on the triage of patients and consideration given to the type of transportation assets available for evacuation.

- Decisions concerning patients left behind. When the patient load exceeds the means to move them, the tactical commander must make the decision as to whether patients are to be left behind. The medical staff officer keeps the tactical commander informed so that he can make timely decisions. Medical personnel and supplies must be left with patients who cannot be evacuated. (Refer to FM 8-10 for additional information.) Every effort will be made to avoid this "solution."

4-3. Combat Health Support for Maneuver and Enabling Operations

a. Choices of Maneuver.

(1) *Penetration.* In this tactic, the attack passes through the threat's principal defensive position, ruptures it, and neutralizes or destroys the threat forces. Of all forms of offensive maneuver, the penetration of main threat defenses normally produces the heaviest patient workload. The commander's plan for the penetration normally has three phases.

- Breaching the enemy's main defensive position.

- Widening the gap created to secure the flanks by enveloping one or both of the newly exposed flanks.

- Seizing the objective with its associated subsequent exploitation.

During breach operations, plans must be in place for clearing casualties off the battlefield since increased number of casualties should be anticipated. Each company team requires an armored ambulance for casualty evacuation. If the engineer company is the breaching force, it must have one armored ambulance in DS. The BAS splints into two treatment teams, Alpha and Bravo, while dividing the two supporting FSMC ambulances. The treatment teams pre-position with the TF prior to LD. Treatment Team Bravo moves forward behind one of the company teams and is designated as the FAS. Treatment Team Alpha (MAS) follows the TF formation. The AXP with a treatment team from the FSMC must be integrated into the TF scheme of maneuver. Without the AXP, the TF medical elements will lose their ability to move as patients collect at the FAS and MAS. The AXP moves forward with the TF combat trains. The TF combat trains should move within 4 kilometers, but no more that 10 kilometers behind the lead elements of the TF. Once the breach is completed, the FAS moves through the breach to the other side, while the MAS moves to the position previously occupied by the FAS. The AXP moves forward to the position previously occupied by the MAS. Patient acquisition starts slowly, but becomes more rapid as the attack progresses. Evacuation routes lengthen as the operation progresses. Heavy preparatory fires which can evoke heavy return fire often precede the penetration maneuver. These threat fires could modify the decision to place evacuation assets as far forward as possible. The FSMC can reinforce the penetration force medical elements. Patient evacuation could be slow and difficult due to a bottleneck at the penetration. Medical evacuation support problems multiply when some combat units remain near the point of original penetration. This is done to hold or widen the gap in threat defenses while the bulk of tactical combat forces exploit or pursue the threat. Treatment elements are placed near each shoulder of the penetration; ground evacuation cannot take place across an avenue of heavy combat traffic. Because of the heavy traffic, the area of the penetration is normally a target for both conventional and NBC weapons. The trigger to push treatment team/BASs through the penetration and where they will go must be identified in the OPORD.

(2) *Envelopment.* In the envelopment, the main or enveloping attack passes around or over the threat's principal defensive positions. The purpose is to seize objectives which cut the threat's escape routes and subject him to destruction in place from flank to rear. Since the envelopment maneuver involves no direct breach of the threat's principal defensive positions, the MEDEVAC system is not confronted with a heavy workload in the opening phase. However, ambulances are positioned well forward in all echelons of CHS to quickly evacuate the patients generated by suddenly occurring contact. Medical treatment facilities moving with their respective formations assist with clearing the battlefield to reduce delays in treatment. After triage and treatment, the patients are evacuated to MTF in the rear by supporting ground ambulances from the FSMC. When the isolated nature of the envelopment maneuver precludes prompt evacuation, the patients are carried forward with the treatment element. This must be planned for in detail and is an extreme measure when no other option is feasible. Expect an increase in mortality from wounds. Again, nonmedical vehicles could be pressed into emergency use for this purpose. When patients must be carried forward with the enveloping forces, CHS commanders use halts at assembly areas and phase lines to arrange combat protection for ground ambulance convoys through unsecured areas. Further, the commander should take advantage of friendly fires and suppression of threat air defenses to call for prearranged air ambulance support missions, or emergency use of medium-lift helicopter backhaul capabilities.

(3) *Infiltration.*

(*a*) Infiltration is a choice of maneuver used during offensive operations. The division can attack after infiltration or use it as a means of obtaining intelligence and harassing the threat. Though it is not restricted to small units or dismounted actions, the division employs these techniques with a portion of its units, in conjunction with offensive operations conducted by the remainder of its units.

(*b*) Combat health support of infiltration is restricted by the amount of medical equipment, supplies, and transportation assets that can be introduced into the attack area. No deployment of BAS treatment teams without their organic transportation should be attempted. Elements of unit-level CHS should be accompanied by their organic vehicles, and ambulances should receive priority for deployment. It may be necessary to man-carry enough BAS equipment into the attack area to provide EMT and ATM; this, however, results in degrading mobility. When the element is committed without its ambulances, patients are evacuated to the BAS by litter bearer teams. These litter teams must be designated and equipped by the commanders in their orders. Noise, light, and litter discipline during evacuation in an infiltration depends on how the casualty was wounded. Disease and nonbattle injury soldiers may not have been noticed by the enemy. If the casualty is a battle injury, the enemy has already detected that element. Once the enemy has detected and engaged the force, causing casualties, maximum allowable use of standard and nonstandard evacuation platforms should be used. This will increase lift capabilities and save time and soldier's lives. Patient evacuation from the BAS and medical resupply of the force may be provided by litter bearers, depending upon distances and degree of secrecy required.

(*c*) When airborne and air assault forces are used, infiltrating elements can land at various points within the threat's rear area and proceed on foot to designated attack positions. As in surface movement, the amount of medical equipment taken could be limited. In airborne operations, the evacuation of patients will be by litter bearers or ground ambulances to collecting points or the BAS and then by FSMC ambulances to the clearing station operated by the FSMC treatment platoon. In air assault operations, the evacuation is by litter bearers to collecting points or the BAS and then by air ambulances to a clearing station. Once the combat element begins the assault on the objective, secrecy is no longer important and its isolated location requires CHS characteristic to airborne and air assault operations until ground linkup.

(4) *Turning movement*. A turning movement is a variation of the envelopment in which the attacking force passes around or over the threat's principal defense positions to secure objectives deep in the threat's rear; thus, forcing the threat to abandon his position or divert major forces to meet the threat. As stated above, the turning movement is a variant to the envelopment in which the attacker attempts to avoid the defense entirely; rather, the attacker seeks to secure key terrain deep in the threat's rear and along his LOC. Faced with a major threat to his rear, the threat is thus "turned" out of his defensive positions and forced to attack rearward at a disadvantage. Medical evacuation support to the turning movement is provided basically in the same manner as to the envelopment. As the operation is conducted in the threat's rear area, LOC and evacuation routes could be unsecured, resulting in delays in resupply and evacuation.

b. *Enabling Operations*.

(1) *Passage of lines*. This situation presents a challenge for the CHS planner. There will be a number of MEDEVAC units using the same air and road networks. Coordination and synchronization are essential if confusion and overevacuation are to be avoided. The medical units of the force manning the line should provide area support to the force passing through. This allows continued mobility for the moving force. The below information facilitates this coordination.

- Radio frequencies and call signs.
- Operation plans and TSOPs.
- Location of MTFs.
- Location of CCP and AXP.
- Main supply route, forward arming and refueling points, and A2C2 data.

(2) *Security operations*. Security operations obtain information about the enemy and provide reaction time, maneuver space, and protection to the main body. Security operations are characterized by aggressive RECON to reduce terrain and enemy unknowns, to gain and maintain contact with the enemy to ensure continuous information, and to provide early and accurate reporting of information to the protected force. See FM 17-95 for definitive information on security operations. The discussion below focuses on how CHS is provided for security operations. Security operations include the following missions: cover, screen, guard, and area security.

(*a*) *Cover*. The covering forces are dependent upon organic resources found in the maneuver battalion medical platoon for initial support. The level of command for the covering force determines the responsibility for the subsequent evacuation plan. In a corps covering force, for example, the corps CHS structure has the responsibility for establishing and operating the MEDEVAC system to support the forward deployed corps forces. This is done to prevent the tactical combat force following the covering forces from becoming overloaded with patients prior to the hand off and passage of lines. The use of CCPs, AXPs, and nonmedical transportation assets (casualty evacuation) to move the wounded is essential. The covering force battle could be extremely violent. Patient loads will be high and the distance to MTFs can be much longer than usual. The effectiveness of the MEDEVAC system depends upon the forward positioning of a number of ground ambulances and the effective integration of corps air ambulances into the evacuation plan.

(*b*) *Screen*. The primary purpose of a screen is to provide early warning to the main body. Screen missions are defensive in nature and largely accomplished by establishing a series of observation posts and conducting patrols to ensure adequate surveillance of the assigned sector. The screen provides the protected force with the least protection of any security mission. Combat health support will be provided by organic medical elements and ambulances teams deployed from the supporting medical company.

(*c*) *Advance, flank, and rear guards*. A guard force accomplishes all the task of a screening force. Additionally, a guard force prevents enemy ground observation of and direct fire against the main body. A guard force reconnoiters, attacks, defends, and delays as necessary to accomplish its mission. A guard force normally operates within the range of main body indirect-fire weapons. A guard force is deployed over a narrower front than a screen to permit concentration of combat power. These forces normally receive MEDEVAC support through the attachment of evacuation teams. The teams evacuate patients to predesignated CCPs along a main axis of advance or to the nearest treatment element providing area support. Employment of air ambulances provides a measure of agility and flexibility.

(*d*) *Area security*. Area security is a form of security that includes RECON and security by designated personnel, airfields, unit convoys, facilities, MSR, LOC, equipment, and critical points. Area security operations are conducted to deny the enemy the ability to influence actions in a specific area or to deny the enemy use of an area for his own purpose. This may entail occupying and establishing a 360-degree perimeter around the area being secured, or taking actions to destroy enemy forces already present. The area to be secured may range from specific points (bridges, defiles) to areas such as terrain features (ridgelines, hills) to large population centers and adjacent areas. Combat health support will be provided by organic and attached medical elements. In area security, Echelon I CHS is provided by organic or attached treatment teams. Echelon II CHS is provided by the supporting medical company via DS and on an area support basis. Depending on the type of area security operations being conducted, both air and ground ambulances may be employed.

(3) *River crossing operations*. The river barrier itself exerts decisive influence on the use of medical units. An attack across a river line creates a CHS delivery problem comparable to that of the amphibious assault. Combat health support elements cross as soon as combat operations permit. Early crossing of treatment elements reduces turnaround time for all crossing equipment that is used to load patients on the far shore. Maximum use of air ambulance assets is made to prevent excessive patient buildup in far shore treatment facilities. Near shore MTFs are placed as far forward as assault operations and protective considerations permit to reduce ambulance shuttle distances from off-loading points. For detailed information on river crossing operations, refer to FM 90-13. Rescuing casualties in the water must be considered by the TF medical planner.

(4) *Reconnaissance operations*. The RECON in force is an *attack* to discover and test the threat's position and strength or to develop other intelligence. The tactical combat force usually probes with multiple combat units of limited size, retaining sufficient reserves to quickly exploit known threat weaknesses. Combat health support techniques follow those discussed above for a movement to contact. Ambulances are positioned well forward and moved at night to enhance secrecy. The echeloning of ambulances is an indication to the threat that an attack is imminent due to the forward placement of CHS. Clearing stations are not established until a significant patient workload develops. Patients received at BASs of reconnoitering units are evacuated to clearing stations as early as practical, or are carried forward with the force until a suitable opportunity for evacuation presents itself. The maximum possible use of air ambulance assets is made to cover extended distances and to overcome potentially unsecured ground evacuation routes.

(5) *Unified action*. The majority of the operations occurring at the present time are joint, interagency, or multinational operations. During the initial planning stages of the operations, the CHS planner must determine who is responsible for providing MEDEVAC support to the force. The CHS planner must also ensure that duplications in support do not exist; that guidelines are established as to eligible beneficiaries; when individuals are to be returned to their own nation's health care delivery system; and what mechanisms exist for reimbursement of services. For additional information, refer to FM 8-42 and Joint Publication 4-02.

(6) *Integrated warfare operations*. Medical evacuation in an NBC environment is discussed in FMs 8-10-6 and 8-10-7.

4-4. Combat Health Support During Night Operations

The battalion surgeon and medical platoon members must anticipate that brigades do a substantial amount of work at night or in limited visibility. They must ensure that the platoon TSOPs (see Appendix F for an example of a TSOP) are available and used throughout the squadron for providing MEDEVAC and treatment at night. Real-life trauma care at night will be enhanced by the ability to use white light (visible light) at the earliest opportunity. Therefore, medical units/elements must establish standard procedures to use white light without compromising the tactical environment. This means training to erect shelters as soon as possible and routinely during hours of darkness. Personnel must understand that some shelters block visible light, but that the same shelters glow when viewed through night vision goggles (NVG). In some extremely mobile situations, ambulance/vehicles could be used to enclose patients and care providers, thus allowing treatment to proceed under white light conditions.

a. Conducting Night Operations. Medical leaders must understand night operations technology and their capabilities for conducting night operations. The brigade, battalion, and squadron surgeons should know how to use both far infrared (IR) devices (and how their capabilities can enhance CHS operations at night) such as the combat identification panel (CIP) and near IR devices, such as the BUDD light and the Phoenix light. See the discussion below on IR and night vision devices. The surgeons need to know the status and amount of equipment on-hand and to identify equipment needed. They must plan the SOPs and METT-TC-specific techniques necessary to perform the CHS mission. For these types of operations, the commander should be advised to consider—

- Appropriating civilian buildings to reduce light and thermal signatures.
- Lightproofing shelters.
- Using nonvisible spectrum light in conjunction with night vision devices.
- Reducing noise signature to a minimum.

In addition, units are susceptible to a night attack. This further slows logistics and CHS activities. Use of chemical lights may be applicable. However, overuse of chemical lights degrades light discipline and security. Chemical lights are visible from a distance of a kilometer or more. Possible techniques for medical units/elements include—

- Chemical lights to light CP areas, thus eliminating generator noise and thermal signature.
- Magnetic holders to allow placement of color chemical lights on vehicles.
- Chemical lights to illuminate areas of vehicle engine compartment for night repairs.
- Chemical light holders to regulate the amount and direction of light.

b. Combat Health Support Considerations.

(1) Light discipline requirements affect CHS operations much as they do supply and maintenance operations. Medical units/elements will use additional fuel to run a vehicle-mounted night site. Treatment operations require lightproof shelters. Patient acquisition is more difficult. Units should employ some sort of casualty-marking system such as luminous tape.

(2) Limited visibility slows MEDEVAC. This requires additional ground ambulances to compensate. In the offense, ambulances move forward with BASs. However, personnel have to accomplish this movement carefully to avoid signaling the threat. Personnel use predesignated AXPs. Medical evacuation by air ambulance is difficult and requires precise grid coordinates as well as prearranged signals and frequencies. As in daylight, CHS operations conducted at night require active participation of all involved units. Operational procedures must include near and far recognition, signaling, predetermined marking of CCPs, routes, and MTFs. Maximum use of modern navigation tools, such as the GPS, IR, and night vision devices, will enhance the ability of medical units/personnel to carry out CHS in support of night missions. Night operating procedures must be routine and practiced as a part of routine operating procedures. This is especially true for medical units/personnel since they have a 24-hour responsibility under all conditions, not just combat operations.

c. Infrared and Night Vision Devices.

(1) A far IR device, such as the CIP, is a "quick fix" device for friendly identification. The thermal taped-covered CIP provides an aid in distinguishing friendly from threat vehicles when thermal sights are used. Combat identification panels do not replace current acquisition, identification, or

engagement procedures. They provide a device visible through thermal sights to increase situational understanding and provide a safety net at the normal engagement range. These devices can be used to further identify medical vehicle and units.

(2) Near IR devices that aid in C2 may be used for signaling and marking devices. The IR beam is an effective means to increase situational understanding and combat effectiveness and improve identification. These devices reduce the fratricide risk when used for marking obstacles, seized terrain, and breached sites. Additionally, these lights are super signaling devices (that is, configuration of certain patterns to indicate unit identification, turn on/off to signal accomplishment of a task, cross a phase line, and signal from one ground position to another specific position, or from ground to air). They are also useful in specialized units such as pathfinders for marking pickup, drop, or landing zones. These are excellent devices for near recognition signaling to guide incoming evacuation vehicles.

(a) BUDD light. The BUDD light operates using active near IR light viewed through image-intensifying devices. These image-intensifying devices are only effective during nighttime conditions. Near IR devices can be directional or omni-directional and emit a steady pulse or codable pulse. The BUDD light is a compact near IR source, using a standard 9-volt (BA-3090) battery as its power source. Both the BUDD light and its power source will fit in the palm of your hand. The average life span of the battery power for a BUDD light is 8 hours of continuous use. The near IR pulse emitted by the BUDD light is similar to a strobe light and pulses every 2 seconds. It is invisible to the naked eye and thermal imagers. The pulse is clearly visible out to 4 kilometers under optimal conditions when pointing the beam directly at the viewer. The directional characteristic of the beam makes it possible to limit observation by an enemy. If used to mark vehicles, care should be taken to minimize the light illuminating the vehicle's surface. The enemy has to have image-intensifying devices to see the lights directly; however, they may see the light being reflected off of vehicles when the lights are employed in a directional mode. This device is most effective for C2 purposes. The BUDD light is also very useful for dismounted operations at night.

(b) Phoenix light. The Phoenix light operates using active near IR light viewed through image-intensifying devices. The Phoenix light can be used as a codable IR beacon. The light is powered by a standard 9-volt (BA-3090) battery. The Phoenix light is ideal for use when positive identification at night must be made out to 4 kilometers under optimal conditions. The IR beacon has a range equal to the BUDD light. One advantage is the ability to code many beacons with different codes (sequence of flashes—including Morse code—up to 4 seconds), enabling anyone to be distinguished in a group. A programmed sequence will repeat until canceled or when the battery expires (same as a BUDD light). Operating instructions include connecting the battery to the Phoenix light. Using a metal object, a coin is best, make connection across the two pins on top of the light. A microminiature red indicator flashes the sequence as the code is entered. At the end of the 4-second memory, a green microminiature indicator will flash, indicating the end of the input sequence. The Phoenix light is now emitting the desired code. To check the code, make a connection across the pins. The green microminiature indicator will flash the code. To change the code, disconnect the battery and repeat the instructions. The Phoenix light also can be used during dismounted operations. The programming of a code can assist in distinguishing one unit from another. An active Phoenix light or BUDD light can be covered or uncovered as necessary to ensure the light is visible only when necessary

(c) *Night vision devices*. There are numerous types of night vision devices in the Army inventory but this subparagraph will focus on what the squadron medical platoon has on its TOE. Each vehicle in the medical platoon will have two night vision devices. The wheeled vehicle driver will use either the AN/PVS-7B (discussed below) or the driver's vision enhancer (DVE). The DVE is a thermal imaging system capable of operating in degraded visibility conditions such as fog, dust, smoke, and darkness. In conditions of reduced visibility, the DVE allows a vehicle to maintain speeds up to 55 to 60 percent of those attained during normal daylight operations. Unlike traditional night vision devices that magnify ambient light, the DVE generates a picture based on very minute variances in temperature in the surrounding environment. It gives the operator visibility to the horizon in total darkness and the ability to recognize a 22-inch object at a distance of 360 feet. It can elevate 35 degrees, depress 5 degrees, and rotate 170 degrees in either direction. The DVE consists of a sensor module, display control module, positioning module, wiring harness, and mounting equipment. A combat DVE and a tactical wheeled vehicle DVE will be available. The ambulance and treatment vehicle drivers will use a DVE if available or continue to wear NVG. The NVG (AN/PVS-7B, a hand-held, head-mounted, or helmet-mounted night vision system) enable walking, driving, weapons firing, short-range surveillance, map reading, treatment of patients, and vehicle maintenance in both moonlight and starlight. It has an IR projector that provides illumination at close ranges and that can be used for signaling. There is a high light-level shutoff if the device is exposed to damaging levels of bright light. There is a compass that attaches to the device and allows for reading an azimuth through the goggles. This device has a weight of 1.5 pounds and operates on two AA batteries. The ambulance and treatment vehicle when fielded will have IR headlights. These IR headlights can be used for assisting drivers who wear NVG and for signaling. As with all lights, extreme caution must be taken in tactical situations. The IR headlights are typically very bright to personnel wearing NVG.

d. *Example Techniques for Using Chemical Lights for Marking and Signaling.*

NOTE

Techniques are only limited to available equipment and imagination. The METT-TC should always take precedence.

(1) For marking, chemical lights can be placed inside standard military short or long pickets to mark routes and positions. The concave side of the picket contains the chemical light and the convex side faces the most likely direction of enemy observation. This technique controls the direction of the light while assisting with such things as MEDEVAC routes, supported unit collection points, AXP, or link-up point identification.

(2) For signaling, tying a chemical light to a length of cord or string and twirling it overhead in a circle is an unmistakable signal. This only needs to be used until recognition (radio) is established. This technique makes use of widely available common supplies. It is especially useful for a unit guiding an incoming ground or air ambulance.

(3) For marking casualty locations, for example, MOUT operations.

(4) For marking triage areas at the BAS to identify patient triage categories.

APPENDIX A

THE GENEVA CONVENTIONS

A-1. Effects of Geneva Conventions on Combat Health Support

The conduct of armed hostilities on land is regulated by both written and unwritten law. This land warfare law is derived from two principal sources—custom and lawmaking treaties such as The Hague and Geneva Conventions. The rights and duties set forth in these conventions are part of the supreme law of the land; a violation of any one of them is a serious offense. The Geneva Conventions are four separate international treaties, signed in 1949, and are respectively entitled:

a. Geneva Convention for the Amelioration of the Conditions of the Wounded and Sick in Armed Forces in the Field (GWS).

b. Geneva Convention for the Amelioration of the Condition of Wounded, Sick, and Shipwrecked Members of Armed Forces at Sea (GWS Sea).

c. Geneva Convention Relative to the Treatment of Prisoners of War (GPW).

d. Geneva Convention Relative to the Protection of Civilian Persons in Time of War (GC).

See FMs 8-10 and 27-10 for definitive information on the Geneva Conventions and the laws associated with land warfare.

A-2. Geneva Wounded and Sick

a. Custodial and medical responsibilities must be carried out for persons (military or civilian) who are wounded as a result of military operations regardless of their nationality or legal status.

NOTE

Persons whose legal status is in doubt are accorded protection and treatment as prisoners of war until their legal status is determined.

b. Collection and treatment of the sick and wounded are responsibilities of medical personnel. The custodial and accounting functions are responsibilities of military police.

A-3. Identification and Protection of Medical Personnel under Geneva Wounded and Sick

Medical personnel who become captured are not considered prisoners of war but retained personnel.

a. Protected personnel include—

(1) Army Medical Department personnel exclusively engaged in the—

- Search for or collection, transport, or treatment of the wounded or sick.

- Prevention of disease.

- Administration of medical units and establishments (for example, this includes personnel such as the office staff, ambulance drivers, cooks, and cleaners that form an integral part of the unit or establishment).

- Veterinary staff functions relating to the administration of medical units and establishments.

(2) Non-AMEDD personnel who have received special medical training, if carrying out their auxiliary medical duties when captured by the enemy are protected. Once in enemy hands, they become prisoners of war when not doing medical work.

(3) Chaplains.

b. Each protected individual must—

(1) Carry a special water-resistant, pocket-sized identity card (DD Form 1934) that—

- Bears the red cross on a white background (the distinctive emblem of the Geneva Conventions).

- Is worded in the national language of the issuing force.

- Contains the surname and first name (at least), date of birth, rank, social security number, protected capacity serving, photograph, signature, and/or fingerprints of carrier.

- Is embossed with the stamp of the appropriate military authority (AR 640-3).

(2) Wear on the left arm a water-resistant armlet bearing the red cross emblem of the Geneva Convention (DA Pamphlet [Pam] 27-1 and FM 27-10).

This paragraph implements STANAG 2454.

A-4. Identification of Medical Units, Facilities, and Vehicles under Geneva Wounded and Sick

a. Identify—

(1) All medical units and facilities except veterinary units. Medical facilities also include the nonpatient care areas, such as those for dining, maintenance, and administration.

(2) Air and surface (ground and water) medical vehicles.

b. How:

(1) Display the distinctive flag of the Geneva Conventions (red cross on a white background) over the unit/facility and in other places on the unit/facility as necessary to adequately identify it. (The other emblem recognized by terms of the Geneva Conventions is the red crescent. Emblems not recognized by the Geneva Conventions but used by other countries, such as the red shield of David by Israel, should also be respected.)

(2) Mark with the distinctive Geneva emblem (red cross on a white background).

(3) The GWS protects from attack any medical vehicle appropriately marked and exclusively employed for the evacuation of the sick and wounded or for the transport of medical personnel and equipment. The GWS prohibits the use of medical vehicles marked with the distinctive emblems for transporting nonmedical troops and equipment.

This paragraph implements STANAG 2931.

A-5. Camouflage of the Geneva Emblem

The NATO STANAG 2931 provides for camouflage of the Geneva emblem on medical facilities where the lack of camouflage might compromise tactical operations. Medical facilities on land, supporting forces of other nations, will display or camouflage the Geneva emblem in accordance with national regulations and procedures. When failure to camouflage would endanger or compromise tactical operations, the camouflage of medical facilities may be ordered by a NATO commander of at least brigade level or equivalent. Such an order is to be temporary and local in nature and countermanded as soon as the circumstances permit. It is not envisaged that large, fixed medical facilities would be camouflaged. The STANAG defines "medical facilities" as "medical units, medical vehicles, and medical aircraft on the ground."

NOTE

Under tactical conditions, the need for concealment may outweigh the needs for recognition (AR 750-1).

A-6. Defense of Self and Patients under Care

a. Protected personnel are—

(1) Authorized to be armed with only individual small arms. (Army Regulation 71-32 provides the doctrine that governs what types of small arms medical personnel are authorized [limited to

pistols or rifles, or authorized substitutes].) These small arms may only be used for defensive purposes. The presence of machine guns, grenade launchers, booby traps, hand grenades, light antitank weapons, or mines in or around a medical unit would seriously jeopardize its entitlement to protected status under the GWS. The deliberate arming of a medical unit with such items could constitute an act harmful to the enemy and cause the medical unit to lose its protected status under the Conventions. This conclusion is not altered in the case of mines regardless of the method by which they are detonated, nor is it altered by the location of the medical unit. If the local non-AMEDD commander situates a medical unit where enemy attacks may imperil its safety, then that commander should provide adequate protection for the medical unit and its personnel.

(2) Permitted to fire only when they or their patients are under direct attack in violation of the GWS. Use of arms by AMEDD personnel for other than protection of themselves or their patients violates the GWS provisions governing the protected status of AMEDD personnel and results in the loss of protected status. Army Regulation 350-41 states the AMEDD personnel and non-AMEDD personnel in medical units will not be required to train or qualify with weapons other than individual or small arms weapons. However, AMEDD personnel attending training at NCO education system courses will receive weapons instruction that is part of the curriculum. This will ensure that successful completion of the course is not jeopardized by failure to attend the weapons training portion of the curriculum.

(3) Responsible for their own defense when operating at locations which preclude their being incorporated within defensive perimeters of nonmedical units. In addition to relying on their special status, medical units can provide for their defense by employing passive defense measures. Passive measures are those taken to reduce the probability of and to minimize the effects of damage caused by hostile action. Examples of these measures are the preparation of individual fighting positions within the immediate unit area; noise and light discipline; posting perimeter sentries; and channeling traffic in the unit area.

b. Protected personnel (under overall security defense plans) will NOT be required—

- To man or help man the perimeter defense of nonmedical units such as unit trains, logistical areas, or base clusters.

- To take offensive action against enemy troops.

- To perform actions that will cause loss of protected status and result in inadequate care of our sick and wounded prisoners of war. The platoon leader must clearly articulate this to all levels of command. The misuse of CHS vehicles/equipment will void all protection granted under the Geneva Conventions.

A-7. Geneva Prisoners of War

a. United States military forces are responsible for EPW from the moment of capture.

b. The echelon commander and medical unit commanders jointly exercise responsibilities for the custody and treatment of the sick, injured, or wounded enemy personnel and detained civilian personnel.

c. The sick, injured, or wounded prisoners are treated and evacuated through normal medical channels but are physically segregated from US and allied patients. Persons other than medical personnel are provided by the echelon commander to guard them. Evacuation of these EPW patients from the combat zone is initiated as soon as their medical conditions permit.

d. When intelligence indicates that large number of EPW may result from an operation, medical units may require reinforcement to support the anticipated additional EPW patient workload. Procedures for estimating the medical workload involved in the treatment and care of EPW patients are described in FM 8-55.

e. Enemy medical personnel are considered retained personnel rather than prisoners of war. They are to be employed to the maximum extent possible in such CHS duties as caring for detained or EPW patients, preferably those of their own armed forces. Captured medical supplies should be used in the care of these patients.

A-8. Geneva Civilian Persons

a. When the US is the occupying power, US forces have the responsibility to ensure that all civilian and refugee subsistence and health service needs are provided.

b. Sick or injured civilian persons resulting from military operations are provided initial medical treatment, as required, in conformance with established theater policies; then, they are transferred to appropriate civil control authorities as soon as possible. When such persons are evacuated, proper accommodations must be provided, including satisfactory conditions of hygiene, health, safety, and nutrition (Articles 49 and 55). In conditions of armed conflict and to the extent practicable, the Army must seek to fulfill the above commitments, as well as to protect and assist civilians and refugees under its control.

A-9. Compliance with the Geneva Conventions

a. As the US is a signatory to the Geneva Conventions, all medical personnel should thoroughly understand the provisions that apply to CHS activities. Violation of these Conventions can result in the loss of the protection afforded by them or prosecution. Medical personnel should inform the tactical commander of the consequences of violating the provisions of these Conventions.

b. The following acts are inconsistent with an individual or facility claiming protected status under the Geneva Conventions:

- Medical personnel are used to man or help man the perimeter of nonmedical facilities, such as unit trains, logistics areas, or base clusters.

- Medical personnel are used to man any offensive-type weapons or weapons systems.

- Medical personnel are ordered to engage enemy forces other than in self-defense or in the defense of patients and MTFs.

- Crew-served weapons are mounted on a medical vehicle.
- Mines or booby traps are placed in and around medical units and facilities.
- Hand grenades, light antitank weapons, grenade launchers, or any weapons other than rifles and pistols are issued to a medical unit or its personnel.
- The site of a medical unit is used as an observation post, a fuel dump, or an ammunition storage site.

c. Possible consequences of violations described in *b* above are—

- Loss of protected status for the medical unit and personnel.
- Medical facilities attacked and destroyed by the enemy.
- Medical personnel being considered prisoners of war rather than retained persons when captured.
- Combat health support capabilities decremented.
- Prosecution for violations of the law of war.

d. Other examples of violations of the Geneva Conventions include—

- Making medical treatment decisions for the wounded and sick on any basis other than medical priority, urgency, or severity of wounds.
- Allowing the interrogation of enemy wounded or sick even though medically not recommended.
- Allowing anyone to mistreat, torture, kill, or in any way harm a wounded or sick enemy soldier.
- Marking nonmedical unit facilities and vehicles with the distinctive emblem, or making any other unlawful use of this emblem.
- Using medical vehicles marked with distinctive Geneva Conventions emblem for transporting nonmedical troops, equipment, and supplies.
- Using a medical vehicle as a TOC.

e. Possible consequences of violations described in *d* above are—

- Criminal prosecution for war crimes.

- Medical personnel being considered prisoners of war rather than retained persons when captured.

NOTE

The use of smoke and obscurants by medical personnel is not a violation of the Geneva Conventions (see FMs 3-50 and 8-10-6 for information on the use of smoke).

APPENDIX B

MANAGEMENT OF INDIVIDUAL HEALTH RECORDS IN THE FIELD

B-1. General

a. This appendix provides guidance on the maintenance of the soldier's individual health record (HREC) and civilian employee medical records (CEMR) in the field. The governing regulation is AR 40-66.

b. Health records are maintained by the MTF that provides primary care for the soldier.

c. Unit commanders will ensure that HRECs are always available to AMEDD personnel who require such records in the performance of their duties. Unit commanders will also ensure that the information in the HRECs is kept private and confidential in accordance with law and regulations governing patient records administration.

d. Health records located at an Echelon I MTFs are maintained by unit medical personnel. The AMEDD officer-in-charge serves as the custodian of the HRECs and CEMRs. Army Medical Department officers are in charge of the HRECs and CEMRs for the members of the units and civilian employees for whom they supply primary medical care. They are also in charge of the HRECs, CEMRs, and the records of other individuals that are receiving treatment from the MTF. Health records are important for the conservation and improvement of the patient's health. Therefore, AMEDD officers will ensure that all pertinent information is promptly entered in the HREC/CEMR in their custody. If any such pertinent information has been omitted, the AMEDD officer will take immediate action to obtain such information from the proper authority and include it in the HREC/CEMR.

B-2. Health Records of Deployed Soldiers

a. Health Records. The HREC (DA Form 3444 or DA Form 8005 Series [Medical and Dental Treatment Records]) of deployed soldiers and the CEMR of deployed civilians will not accompany them to the combat area.

(1) The supporting MTF will initiate a DD Form 2766 (Adult Preventive and Chronic Care Flowsheet), DD Form 2766C (Adult Prevention and Chronic Care Flowsheet [Continuation Sheet]), DD Form 2795 (Pre-Deployment Health Assessment Questionnaire), and DD Form 2796 (Post-Deployment Health Assessment Questionnaire). If an individual deploys, the DD Form 2766 and DD Form 2766C will be photocopied prior to deployment and the copy will be kept in the medical record. The original DD Form 2766 and any DD Forms 2766C will accompany the individual to the field. The DD Form 2766 serves as the treatment folder for the individual that is deployed; other forms, such as DD Form 2766C, DD Form 2795, DD Form 2796, and Standard Form (SF) Form 600 (Health Record—Chronological Record of Medical Care) will be filed on the fastener inside DD Form 2766. The photocopies of the DD Form 2766 and DD Form 2766C will be removed and shredded when the originals are placed back into the HREC or CEMR. Forms that had been filed inside the DD Form 2766 folder will be removed and place in the HREC or CEMR (in the regular treatment folder).

(2) When processing individuals for deployment, the MTF and dental treatment facility (DTF) will audit each individual's HREC or CEMR and record essential health and dental care information on DD Form 2766. If a HREC or CEMR is not available, DD Form 2766 will be completed based on individual interviews and any other locally available data. A HREC may not be available for most Individual Ready Reserve, Individual Mobilization Augmentees, and retired personnel because these HREC may remain on file at the Army Reserve-Personnel Command (AR-PERSCOM) or the Department of Veterans Affairs.

(3) Upon notification of deployment, all military personnel will complete DD Form 2795.

(*a*) The individual being screened will fill out the section entitled *Demographics* on page 1, and the section entitled *Health Assessment* on page 2. These sections are self-explanatory.

(*b*) The health assessment administrator will fill out the boxed area on page 1 entitled *Administrator Use Only*, and will answer the user's questions on filling out the form. The administrator will document the deployment location as well as the completion date of the pre-deployment evaluation on DD Form 2766, Block 11—*Pre-/Post-Deployment History*. This does not apply to classified operations.

(*c*) The health care provider will fill out the section entitled *Pre-Deployment Health Provider Review* on page 2.

(*d*) A copy of the form will be filed on the fastener inside the DD Form 2766 folder; one copy will remain in the HREC, and the original form will be sent to the Army Medical Surveillance Activity, ATTN: MCHB-TS-EDM/Deployment Surveillance, Building T-20, Room 213, 6825 16th Street NW, Washington, DC 20307-5000.

(4) Department of Defense Directive (DODD) 6490.2 and Department of Defense Instructions (DODI) 6490.3 state that to the extent applicable, medical surveillance activities will include essential DOD civilian and contractor personnel directly supporting deployed forces, consistent with plans established under DODI 1400.32 and DODI 3020.37. If DD Form 2795 is used for civilians, a copy of the form will be filed on the fastener inside the DD Form 2766 folder; one copy will remain in the CEMR, and the original form will be sent to the Army Medical Surveillance Activity.

(5) If the deployed individual is taking part in a classified operation, the pre-deployment evaluation (DD Form 2795) is still required, but the form will be maintained only in the personnel folder.

(6) The completed DD Form 2766 and a copy of any printout from an automated immunization tracking system will be provided to the individual's command, or to the individual if he or she is an individual replacement, and then handed off to the MTF in the AO responsible for providing primary medical care to that individual. That MTF will maintain the DD Form 2766 as an outpatient field file for reference as needed. The MTF will ensure that the ABO/Rh blood type from a verified blood bank typing is recorded in Block 10. The field file will consist of, in part, DD Form 2766, DD Form 2795, and possibly DD Form 2766C, DD Form 2796, SF 600, SF 558 (Medical Record—Emergency Care and Treatment), SF 603 (Health Record—Dental), or DD Form 1380. These forms will be filed on the fastener inside the DD Form 2766. For detailed information on how to complete the DD Form 1380, see Appendix C, FM 8-10-6.

(7) If DD Form 2766 is not available, the individual's field file may be managed as a "drop" file (forms not attached) and integrated into the DD Form 2766 when it is available.

b. Forwarded Deployed Force. If time permits, follow guidance in *a*(1), (2), and (3) above. If not, consolidate HREC in-country and process when time permits.

c. Limited Contingency Operations. Retain the HREC at the MTF and DTF providing primary care. If the servicing primary care facility closes, forward the HREC to the MTF or DTF indicated by the servicing medical department activity (MEDDAC) and dental activity. If full mobilization occurs, follow guidance in *a*(1), (2), and (3) above.

d. Units That Do Not Process Through a Mobilization Station. Units that do not process through a mobilization station before deployment or otherwise do not have access to an MTF or DTF will follow the procedures in *b* above.

B-3. Use of Field Files/DD Form 2766

a. If a member's primary MTF changes, the field file/DD Form 2766 should be moved to the gaining MTF.

b. If a member requires admission to the hospital, every attempt will be made to forward the field file/DD Form 2766. The file will be returned to the member's primary MTF if disposition is RTD.

B-4. Storage of Health Records and Civilian Employee Medical Records

Forward deployed (Echelon I and Echelon II) MTFs will secure field chest or field file containers in quantities sufficient for the troop and civilian employee population supported. They will maintain the DD Form 2766 for each individual receiving primary medical care from their MTF.

B-5. Establishment and Management of the Field File in the Operational Area

a. A DD Form 2766 and the medical records identified above will be maintained by medical companies operating an Echelon II MTF or the medical platoon/section that operates an Echelon I MTF, or will be handed off to the MTF providing their primary care.

b. Supported units will be required to provide the primary care MTF a battle roster of personnel assigned. This roster should be provided when personnel assignment changes are made or upon request.

c. The MTF, when possible, will attempt to ensure that the HREC or CEMR accompanies the medically evacuated individual.

d. If an individual's primary MTF changes, the HREC or CEMR will be transferred to the gaining MTF.

e. If an individual requires hospital admission, every attempt will be made to forward the HREC or CEMR to the admitting hospital.

f. When the MTF determines that an individual was evacuated without the DD Form 2766 and other medical records in the file, then the individual's DD Form 2766 and other medical records are forwarded to the medical C2 headquarters responsible for regulating patients out of the AO. The medical C2 headquarters forwards the outpatient field file to the hospital where the patient was evacuated. The hospital patient administration section will attach the file to the inpatient chart and the file is evacuated with the patient out of the AO or theater.

B-6. Health Assessments after Deployment

a. All military personnel will complete DD Form 2796 prior to leaving the AO.

(1) The individual being screened will fill out the section entitled *Demographics* on page 1 and the section entitled *Health Assessment* on page 2. These sections are self-explanatory.

(2) The health assessment administrator will fill out the boxed area on page 1 entitled *Administrator Use Only* and will answer the user's questions on filling out the form. The administrator will document the deployment location (if this information is missing) and the completion date of the post-deployment evaluation on DD Form 2766, Block 11—*Pre-/Post-Deployment History*. This does not apply to classified operations.

(3) The health care provider will fill out the section entitled *Post-Deployment Health Provider Review* on page 2.

b. If a situation does not allow this health screening prior to departure, the individual's commander will ensure that the health assessment is completed and submitted to the local MTF commander within 30 days of the individual's return. The local MTF commander will ensure that a procedure is in place for submitting the original DD Form 2796 to the Army Medical Surveillance Activity and for filing a copy in the HREC.

c. If the DD Form 2796 is completed prior to leaving the AO, a copy of the form will be filed in the DD Form 2766 folder until it can be integrated into the HREC. The original DD Form 2796 will be submitted to the Army Medical Surveillance Activity, ATTN: MCHB-TS-EDM/Deployment Surveillance, Building T-20, Room 213, 6825 16th Street NW, Washington, DC 20307-5000.

d. The post-deployment assessment of Reserve Component personnel must be completed prior to release from active duty if not completed before redeployment. Reserve Component personnel who have been deployed will also complete DD Form 2697 (Report of Medical Assessment) according to AR 40-501. Reserve Component personnel who are called to active duty but never actually deployed will only complete DD Form 2697.

e. If DD Form 2796 is used for civilians, the form will be completed prior to leaving the AO. If a situation does not allow this health screening prior to departure, the individual's commander will ensure that the health assessment is completed within 30 days of the individual's return. If the DD Form 2796 is completed prior to leaving the AO, a copy of the form will be filed in the DD Form 2766 folder until it can be integrated into the CEMR. The local commander will ensure that a procedure is in place for submitting the original DD Form 2796 to the Army Medical Surveillance Activity and for filing the copy in the CEMR.

f. If the deployed individual is taking part in a classified operation, the post-deployment evaluation (DD Form 2796) is still required, but the form will be maintained only in the personnel folder.

B-7. Field Record Administration after Hostilities Cease

a. Field files/DD Form 2766 will be integrated with the HREC or CEMR after demobilization at the home station or at mobilization stations.

(1) On return to the MTF (post-deployment), forms, such as SF 600, will be removed from the DD Form 2766 folder and placed with the other SF 600 in the medical record.

(2) DD Form 2795 and DD Form 2796 will be removed from the DD Form 2766 folder and placed as shown in Figures 5-1, 5-2, or 7-1 of AR 40-66. If a previously photocopied DD Form 2795 is contained in the record, only one of the DD Forms 2795 will be kept; the other will be removed and shredded.

(3) The photocopies of the DD Form 2766 and DD Form 2766C will also be removed and shredded when the originals are placed back into the record. Field files/DD Form 2766 will be forwarded to AR-PERSCOM for those members whose HREC is maintained at AR-PERSCOM.

b. Each continental United States (CONUS) MTF must request records from AR-PERSCOM for those members who remain on active duty and are assigned for support upon demobilization.

c. Field files will be integrated with the HREC maintained at home station or mobilization station. Field files will be forwarded to Army Reserve Personnel Center (ARPERCEN) for those members whose HREC is maintained at ARPERCEN.

d. Each CONUS MTF must request records from ARPERCEN for those soldiers who remain on active duty and are assigned for support upon demobilization.

APPENDIX C

COMBAT LIFESAVER

C-1. Role of the Combat Lifesaver

a. Immediate far-forward first aid is essential on a widely dispersed and fluid battlefield to prevent soldiers from dying of wounds. Medical personnel may not be able to reach and apply EMT to all wounded soldiers at all points on the battlefield in a timely manner. The CLS is a nonmedical soldier trained to provide advanced first aid/lifesaving procedures beyond the level of self-aid or buddy aid. The CLS is not intended to take the place of medical personnel, but to slow deterioration of a wounded soldier's condition until medical personnel arrive. Functioning as a CLS for the soldier is a secondary mission undertaken only when the tactical situation permits. Even though this is secondary to his primary mission, the CLS has proven to be very effective in saving wounded soldiers lives. The CLS program is implemented according to AR 350-41.

b. The AirLand Battle doctrine was developed for a widely dispersed, rapidly moving battlefield. As was determined for the AirLand Battle doctrine and applicable to the Force XXI doctrine, there are constraints on the rapid acquisition of casualties and medical treatment. These constraints in many cases limit the ability of medical personnel to provide immediate, far-forward medical treatment. The plan developed to provide care for soldiers under AirLand Battlefield doctrine included the CLS. Under Force XXI doctrine, the CLS is an intricate part of providing care for the wounded soldier.

c. The CLS is a bridge between the self-aid/buddy aid training provided all soldiers and the medical training given to the trauma specialist. The CLS is given additional first aid training and training in selected medical tasks (such as initiating an intravenous infusion and providing limited care to a soldier suffering from BF).

C-2. Training the Combat Lifesaver

A correspondence course has been developed for training both active duty and Reserve Component personnel. The course is offered only in a group study mode and with training taking place at the unit level. Classroom instruction is provided by qualified instructors selected by the battalion commander or battalion/ squadron surgeon. Testing is performed at the unit level using the written and performance tests furnished in the correspondence course. Training and testing will be conducted according to the tasks, conditions, and standards established by the Academy of Health Sciences, AMEDDC&S, and published in correspond-ence course training materials. Students who successfully complete the written and performance tests will receive promotion points and be certified as a CLS. The course consists of student subcourse texts, student examination, and an instructor's manual.

C-3. Administering the Combat Lifesaver Course

a. Equipment and Supplies. Arrange for equipment and supplies as early as possible. The purchase of some items, such as intravenous infusion trainers and rescue breathing mannequins, may be required. The local Training and Audiovisual Support Center may have these items available. Training items will not be provided by either the Army Institute for Professional Development (AIPD) or the Academy of Health Sciences, AMEDDC&S.

b. Enrollment. Enrollment request should be sent to AIPD at least 6 weeks prior to beginning of the course according to instructions found in DA Pam 350-59. To establish this group enrollment, AIPD must receive the following:

(1) A request for training should be signed by the battalion commander or an LTC or higher and identify the primary instructor (group leader). All instructors of the Combat Lifesaver Course (Course Number: 081 F11) must meet the following criteria: hold a primary of MOS 91W or 18D, or be a licensed paramedic (state or national), registered nurse, PA, or physician. These requirements reflect the level of expertise necessary to resolve medical emergencies associated with the tasks to be taught.

(2) One DA Form 145 (Army Correspondence Course Enrollment Application) enrolling the primary instructor in IS0826, along with a list of assistant instructors, if any, is included. The DA Form 145 should be signed by a responsible official who has the authority to requisition or acquire the necessary medical supplies to support the training.

(3) A roster of students enrolling in IS0824 and IS0825 is provided. The roster must include each student's full name, rank, social security number, and component code. Ideally, no more than 15 to 20 students should be assigned to each group.

c. Facilities. Facilities must be reserved well in advance. The facilities chosen should allow clear observation of demonstrations and provide room for student practice. Handwashing devices are required.

d. Course Material. The course consists of two subcourses (shipped in one box), 40 credit hours consisting of self-study materials and approximately 3 days of classroom instruction and testing materials. Testing includes both proctored multiple-choice and performance examinations. Only one examination response (IS0827) is returned to AIPD for grading upon completion of IS0824 and IS0825. This is a GO or NO/GO course.

e. Recertification. Combat lifesaver tasks are perishable skills. Combat lifesavers must be recertified every 12 months on the performance-tested tasks in the CLS course. The printed material in the correspondence course can be locally reproduced and used for sustainment training. Do not reenroll soldiers requiring recertification. Unit instructors can accomplish this. The instructor's guide also includes a test appropriate to recertification. It is the responsibility of the S1, not the medical platoon leader, to ensure that personnel matters concerning the CLS program are resolved.

f. Program Managers. Program managers are not authorized to augment correspondence course material, change the length of the course, or increase or delete items contained in the CLS aid bag.

g. Sustainment Training. To the extent needed to sustain skill proficiency, CLS will be exercised during home station training activities (to include field-training exercises) and during deployment for training (to include rotations through combat training centers).

h. Additional Information. For additional information, write to Student Service Division at the Army Institute for Professional Development, US Army Training Support Center, Newport News, VA 23628-0001, or call DSN 927-3322/2127 or commercial 757-878-3322/2127.

C-4. Combat Lifesaver Aid Bag

a. Each certified CLS will be issued a CLS aid bag. The aid bag will be packed in accordance with the prescribed packing list and will be secured as a sensitive item (for example, weapon or night vision devices) at unit level. The aid bags will be issued to the CLS only upon deployment (training and actual).

b. It is the responsibility of each CLS to ensure that—

(1) His aid bag is stocked according to the prescribed packing list.

(2) All stocked items are serviceable.

(3) Items have not exceeded their expiration dates.

C-5. Class VIII Resupply of Combat Lifesaver Aid Bag

a. During garrison operations, Class VIII resupply of the CLS aid bags is conducted in the following manner:

(1) Units with assigned medical personnel will request Class VIII resupply for CLS aid bags.

(2) Divisional, brigade, and regimental units without assigned medical personnel request Class VIII support through the DMSO, brigade medical supply section, or regimental medical supply section. Force XXI units will request Class VIII resupply from either the DSMC medical supply section or the FSMC medical supply section.

(3) Nondivisional units request Class VIII support from the installation medical supply activity, which is normally the medical supply office of the medical department activity or a US Army medical center.

b. During field operations, resupply is accomplished in the same manner as during garrison operations with two exceptions.

(1) Nondivisional units obtain resupply support from the nearest medical unit available.

(2) Emergency resupply is provided to all units by the nearest medical unit capable of doing so.

c. Field medical units maintain stocks of medical material with which to effect resupply of CLS aid bags during field operations.

APPENDIX D

INTERIM BRIGADE COMBAT TEAM

Section I. INFANTRY BATTALION MEDICAL PLATOON

D-1. Medical Platoon

The medical platoon of the infantry battalion is organized with a headquarters section, a treatment squad, an evacuation section, and a combat medic section (see Figure D-1). For more detailed information on the functions and operations of the medical platoon, see Chapter 2, Sections IV and VI.

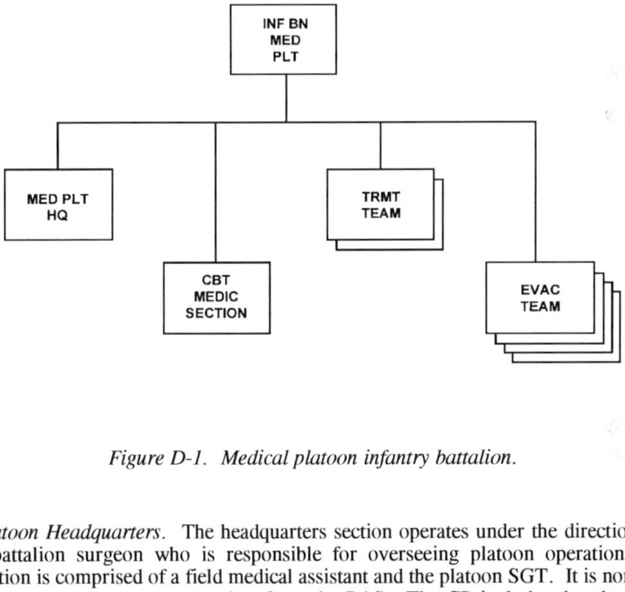

Figure D-1. Medical platoon infantry battalion.

a. Platoon Headquarters. The headquarters section operates under the direction of the medical platoon leader/battalion surgeon who is responsible for overseeing platoon operations. The platoon headquarters section is comprised of a field medical assistant and the platoon SGT. It is normally collocated with a treatment team or a treatment squad to form the BAS. The CP includes the plans and operations functions performed by the field medical assistant. The platoon has access to the infantry battalion HHC wire communication network for communications with all major elements of the battalion. Wireless communications for this section consists of a tactical FM radio mounted in the platoon headquarters vehicle. The medical platoon employs an FM radio network for CHS operations. The platoon headquarters section serves as the NCS for the platoon (see Figure D-2). Table D-1 lists the information and communications assets available to the platoon.

(1) The field medical assistant, an MS officer, is the operations/readiness officer for the platoon. He is the principal assistant to the platoon leader for operations, administration, and logistics. The field medical assistant coordinates CHS operations with the infantry battalion S1 and S4, and MEDEVAC with the brigade support medical company (BSMC).

(2) The platoon SGT assists in supervising the operations of the platoon. He also serves as the evacuation section SGT.

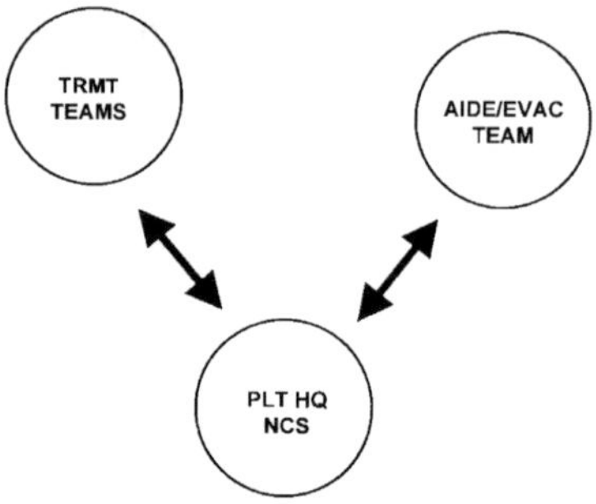

Figure D-2. Medical platoon internal communications net.

Table D-1. Information and Communications Assets Available to the Infantry Battalion Medical Platoon

RADIO SETS	
AN/VRC-89F	PLATOON HEADQUARTERS VEHICLE AND TREATMENT TEAM ALPHA (SURGEON) VEHICLE (1 EACH)
AN/VRC-88F	TREATMENT TEAM BRAVO
AN/VRC-90F	PLATOON AMBULANCE (1 EACH)
ROUTERS	**COMPUTER SYSTEMS**
TACTICAL LAN LAN ROUTER	MC4 LAPTOPS FOR EACH TREATMENT AND AMBULANCE VEHICLE
	MC4 (DISMOUNTED, HAND-HELD)
	PIC
	READER/WRITER (1 EACH, TRAUMA SPECIALIST)
	FBCB2, 1 IN EACH VEHICLE ASSIGNED TO THE MEDICAL PLATOON
	OTHER SYSTEMS
FBCB2	MEDICAL PLATOON VEHICLE (1 EACH)
GPS	MEDICAL PLATOON VEHICLE (1 EACH)
EPLR	MEDICAL PLATOON VEHICLE (1 EACH)
BCIS	MEDICAL PLATOON VEHICLE (1 EACH)
DVE	MEDICAL PLATOON VEHICLE (1 EACH)

b. Treatment Squad. The treatment squad consists of two treatment teams (Teams Alpha and Bravo). They operate the BAS and provide Echelon I medical care and treatment. This includes sick call, EMT, and ATM. Team Alpha is staffed with an operational medicine officer (primary care physician/ battalion surgeon), a health care SGT, and two health care specialists. Team Bravo is staffed with a PA, a health care SGT, and two health care specialists. The physician, PA, and health care SGT and specialists are trained to provide EMT and assist with ATM procedures, commensurate with their occupational specialties. The treatment teams can operate for limited times in split-based operations in DS of battalion units. The teams can also operate in split-based operations when the BAS must move to a new location. One team remains at current location and continues to treat patients while the other team moves to the new location and establishes patient care capabilities. Once the jump team has established a treatment capability at the new location, the other team evacuates or returns to duty all patients and moves to the new location.

(1) The medical platoon leader is a working physician on Treatment Team Alpha. He is also the battalion surgeon. In this role, he is a special staff officer and advisor to the battalion commander on employment of the medical platoon and on the health of the battalion. He is also the supervising physician (field surgeon) of the medical platoon's treatment squad. This officer is responsible for all CHS provided by the platoon. His responsibilities include—

- Planning and directing CHS for the infantry battalion. He does this in conjunction with the battalion S1, who is the coordinating staff officer responsible to the commander for health and welfare of the troops.
- Advising the infantry battalion commander and his staff on CHS operations and the medical threat.
- Supervising the administration, discipline, maintenance of equipment, supply functions, organizational training, and employment of medical platoon personnel.
- Examining, diagnosing, treating, and prescribing courses of treatment for patients, to include ATM.
- Training CLS.
- Supervising the battalion CSC program.
- Planning and conducting humanitarian assistance programs, when directed.
- Providing PVNTMED support for the battalion. Requesting PVNTMED support from the brigade for PVNTMED requirements beyond his capabilities.
- Planning and overseeing PVNTMED training for battalion personnel.
- Advising the commander on the health of the battalion.
- Supervising the training of unit field sanitation teams.

(2) The PA performs patient health care and administrative duties. The PA is ATM-qualified and works under the clinical supervision of the medical officer. The PA assumes the duties of the battalion surgeon/medical platoon leader in his absence. He performs the following duties:

- Establishes and conducts treatment team operations when deployed in split-based operations.

- Treats, within his ability, sick or injured patients. He refers those patients requiring treatment beyond his capability to the supervising physician.

- Provides EMT and ATM for wounded and DNBI patients.

- Conducts training for battalion personnel in first aid procedures (self-aid, buddy aid, and CLS), field sanitation, evacuation of the sick and wounded, and the medical aspects of injury prevention.

- Assists in the conduct of the battalion CSC program, to include individual and leader training on the prevention of BF and other stress-related conditions.

- Trains medical personnel in EMT procedures.

- Advising the command on PVNTMED concerns and conducting PVNTMED activities within the capabilities of medical platoon personnel. Assist in training unit field sanitation teams.

c. Combat Medic Section. Trauma specialists are allocated to the companies of the infantry battalion on the basis of one trauma specialist per platoon. The platoon trauma specialist normally locates with, or near, the platoon leader or platoon SGT. When the platoon is moving on foot in the platoon column formation, he positions himself near the element leader trailing the base squad forward of the second team. When the platoon is mounted, the trauma specialist will normally ride in the same vehicle as the platoon SGT. A health care SGT is allocated on the basis of one per infantry company. The company health care SGT normally collocates with the 1SG. When the company is engaged, he remains with the 1SG and provides medical advice as necessary. As the tactical situation allows, he will manage the company CCP, provide medical treatment, and prepare patients for MEDEVAC.

d. Evacuation Section. Medical platoon ambulances provide evacuation and en route care from the soldier's point of injury or the company's CCP to the BAS. The ambulance team supporting the company works in coordination with the trauma specialists supporting the platoons. When a casualty occurs in a fighting vehicle, the evacuation team will move as close to the vehicle as possible, making full use of cover, concealment, and defilade. Assisted, if possible by the vehicle's crew, they will extract the casualty from the vehicle and administer EMT. In mass casualty situations, nonmedical vehicles may be used to assist in casualty evacuation as directed by the supported commander. Plans for the use of nonmedical vehicles to perform casualty evacuation should be included in the infantry battalion's TSOP and OPORD. Patients are evacuated from the BAS to the BSMC by BSMC ground ambulances or FSMT aeromedical evacuation aircraft. During entry operations, based on the current concept, air ambulances will not be available for the first 96 hours.

D-2. Medical Communications for Combat Casualty Care

a. The MC4 system will be a theater, automated CHS system, which will link commanders, health care providers, and supporting elements, at all echelons, with integrated medical information. The battalion will have one MC4 (dismounted) with two laptops and one server. The system will provide digital enablers to connect, both vertically and horizontally, all ten CHS functional areas. When developed, the MC4 system will receive, store, process, transmit, and report medical C2, medical surveillance, casualty movement/tracking, medical treatment, medical understanding, and CHL data across all echelons of care. This will be achieved through the integration of a suite of medical information systems linked through the Army data telecommunications architecture. The MC4 system begins with the individual soldier and continues throughout the health care continuum. The best way to visualize the MC4 system capability is as a piece of the Army digital computer network where all ten CHS functional areas have been digitized and this CHS information is available to specified commands, supported units, and their personnel. When fully developed, not only will the MC4 system provide Army commanders with CHS information, but will provide them with a seamless transition to the joint health service support environment.

b. The MC4 system will consist of three basic components: software, hardware, and telecommunications systems.

(1) *Software system.*

(*a*) The joint TMIP will provide GOTS/COTS software and interoperability standards to support joint theater operations. The software provides an integrated medical information capability that will support all echelons of care in a theater of operations with links to the sustaining base. Medical capabilities provided by the software to support commanders in the theater will address—medical C2 (including medical capability assessment, sustainability analysis, and medical intelligence); CHL (including blood product management and medical maintenance management); patient evacuation; and health care delivery.

(*b*) The MC4 system will support Army-unique requirements and any software needed to interface with Army information systems such as CSSCS, GCSS-A, FBCB2, Warrior Programs, and the Movement Tracking System.

(2) *Hardware systems.* The hardware will consist of COTS automation equipment supporting the above software capabilities. Examples include, but are not limited to, computers, printers, networking devices, and the PIC.

(3) *Telecommunications systems.* The MC4 system will rely on current and proposed Army solutions for tactical, operational, and strategic telecommunications systems to transmit and receive digitized medical information throughout the theater and back to the sustaining base. There will be no separate AMEDD communication system. Telecommunications at brigade and below will be accomplished through the tactical internet; above brigade level, telecommunications will be accomplished through the WIN architecture. The MC4 system will include hardware or software required to interface with current and emerging technologies supporting manual, wired, and wireless data transmission. At end-state, the MC4 system users will exchange data electronically via the WIN architecture. In the interim, commercial

satellite and/or high frequency radio will be fielded to selected medical units (for example, MDT and so forth) receiving the MC4 system to support high bandwidth requirements until the WIN architecture is fully fielded. Personnel operating satellite assets are resourced in the MDT TOE and will be located with the MDT.

(4) *Patient treatment recording system.* In the future under MC4, medical information about each soldier will be entered into a local database maintained at the supporting BAS or troop medical clinic. This information will include the soldier's immunization status, medical deployability status, and dental deployability status. Until a digital patient record and the PIC are fully functional and fielded, and in accordance with AR 40-66, a field medical record jacket, DD Form 2766, and its accompanying records will be maintained by the soldier's primary care provider. See Appendix B for definitive information on management of the individual health records in the field.

c. When fielded, the MC4 system will be employed by the medical platoon in the conduct of its mission. The medical platoon will employ FBCB2 and other communications enablers for preparing and submitting daily patient feeder reports, transmitting medical surveillance information, requesting supplies, and conducting other administrative activities (see Table D-1 for types of medical platoon communication systems and enablers).

- The medical platoon headquarters will employ an FM radio as the platoon NCS.

- The treatment squad will employ FM radios and notebook computers to conduct treatment team/squad communications. The notebook computers will be used to read and enter patient data on the PIC; to conduct teleconsultation (TCON) and TMEN activities; provide patient flow/disposition information; and DNBI information to the battalion S1 and higher echelon medical leadership. The physician will provide TCON to the PA and receive TCON from the medical company or other medical activities out of theater through the MDT. The notebooks will be linked to others through FM radios. The platoon physician and PA will provide TMEN for ambulance personnel.

- The evacuation section will employ FM radios and notebook computers to conduct MEDEVAC communications. The notebook computers will be used to read and enter patient care information on the PIC. They will also use this device to receive TMEN from the physician and PA. The notebooks will be linked to the FM radios for communications operations.

- The trauma specialists will employ a hand-held device for reading and entering patient care information on the PIC.

D-3. Battalion Combat Health Support Planning

a. *Planning Considerations.* The battalion surgeon assisted by the field medical assistant and the platoon SGT is responsible for the CHS plan for the infantry battalion. As operational requirements or the mission changes, the CHS plan must be updated. The following factors should be considered:

- Commander's information requirements.

- Results of the mission analysis.
- Commander's intent.
- Planning guidance.
- Courses of actions.
- Tactical plan.
- Enemy.
- Terrain.
- Troops supported (unit and attached).
- Weather.
- Threat (including medical threat).
- Operational conditions and constraints.
- Civilian populace in the AO.
- Medical personnel status.
- Equipment status of the medical platoon.
- Supply status including Class VIII.
- Communications capability.
- Nuclear, biological, and chemical defense (including radiation OEG).
- Patient decontamination.
- Medical platoon's training status.
- Unit field sanitation team training.
- Unit personnel first aid training status, to include CLS.
- Casualty estimates.
- Medical evacuation requirements and capabilities.

- Nonmedical support requirements from the battalion.
- Area support requirements.
- Special operations support requirement.
- Mass casualty operations in accordance with the TSOP.
- Medical records and reports requirements.
- Policy and procedure updates.

The foundation of the battalion CHS plan is the battalion commander's guidance and the brigade CHS plan.

b. *Plans*. See Chapter 3, paragraph 3-1.

c. *Operation Orders*. See Chapter 3, paragraph 3-2.

d. *Rehearsals*. See Chapter 3, paragraph 3-3.

D-4. Combat Health Support During Night Operations

See Chapter 4, paragraph 4-4.

D-5. Combat Health Support Tactical Standing Operating Procedures

The battalion surgeon/medical platoon leader is responsible for the development of the CHS annex for the battalion TSOP. The purpose of a TSOP is to establish routine protocols. The TSOP should not be dependent upon METT-TC factors. If a specific decision is required each time, it should not be included in the TSOP. The battalion TSOP is based on its higher headquarters TSOP and serves as the foundation for subordinate units to develop their TSOP. The battalion CHS annex to the TSOP should be clear and concise, yet provide sufficient detail of any procedural requirements. The CHS annex to the TSOP must reflect procedural guidance that supports current mission and doctrinal requirements. The CHS annex to the battalion TSOP should be maintained and reviewed at least every 6 months and revised as required. Most importantly, the TSOP must be used during training and understood at all levels prior to deployment or it has no real value.

Section II. RECONNAISSANCE, SURVEILLANCE, AND TARGET ACQUISITION SQUADRON

D-6. Medical Platoon

The medical platoon of the RSTA squadron provides Echelon I CHS for the squadron. The platoon is organized with a headquarters section, a treatment squad and an evacuation section (see Figure D-3). For more detailed information on the functions and operations of the medical platoon, see Chapter 4.

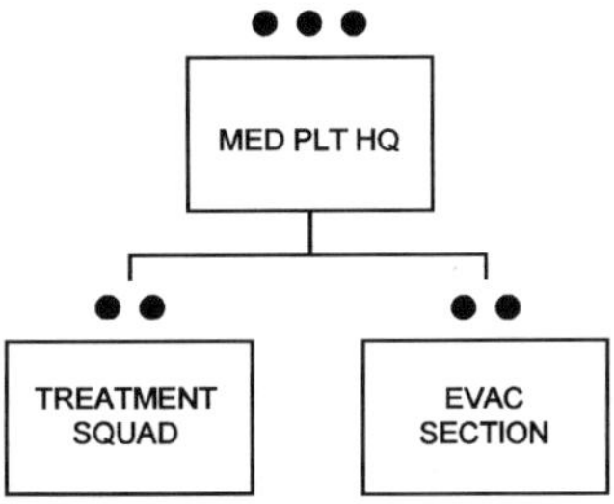

Figure D-3. Medical platoon, reconnaissance, surveillance, and target acquisition squadron.

D-7. Platoon Headquarters

a. The platoon headquarters section, under the direction of the platoon leader/surgeon that is assigned to the treatment squad, provides the C3 and logistics for the platoon. The platoon headquarters section is comprised of a field medical assistant, the platoon SGT and a medical specialist. It is normally collocated with a treatment team/squad to form the squadron aid station. The CP includes the plans and operations functions performed by the field medical assistant. The platoon has access to the HHT and the supported troop wire communication network for communications with all major elements of the squadron. Wireless communications for this section consists of a tactical FM radio mounted in the platoon headquarters vehicle. The medical platoon employs an FM radio network for CHS operations. The platoon headquarters section serves as the NCS for the platoon. Table D-1 lists the information and communications assets available to the platoon.

b. The medical platoon leader is a working physician on Treatment Team Alpha. He is the medical advisor to the squadron commander and his staff. He is also the supervising physician (field surgeon) of the medical platoon's treatment teams. This officer is responsible for all medical treatment provided by the platoon. His responsibilities include—

- Planning and directing CHS for the RSTA squadron.

- Advising the squadron commander and his staff on CHS operations and the medical threat.

- Supervising the administration, discipline, maintenance of equipment, supply functions, organizational training, and employment of assigned or attached personnel.

- Examining, diagnosing, treating, and prescribing courses of treatment for patients, to include ATM.

- Training the squadron's CLS.

- Supervising the squadron's CSC program, to include individual and leader training on the prevention of BF and other stress-related conditions.

- Planning and conducting humanitarian assistance programs when directed.

- Overseeing and coordinating the MEDEVAC of patients.

c. The field medical assistant, an MS officer, is the operations/readiness officer for the platoon. He is the principal assistant to the platoon leader for operations, administration, and logistics. The field medical assistant coordinates CHS operations within the squadron and with supporting brigade or corps medical elements. He coordinates MEDEVAC with the BSMC. The platoon SGT assists in supervising the operations of the platoon. He also serves as the evacuation section SGT.

D-8. Treatment Squad

The treatment squad has two treatment teams (Teams Alpha and Bravo) that are the basic medical treatment elements of the squadron aid station. They provide Echelon I medical care and treatment. This includes sick call, EMT, ATM, and triage for the management of mass casualty situations. Team Alpha is staffed with an operational medicine officer (primary care physician/squadron surgeon), a treatment squad leader (E-6), a health care SGT (E-5), and one health care specialist (E-3). Team Bravo is staffed with a PA, a health care SGT, and two health care specialists. The physician and PA are trained in ATM procedures, commensurate with their occupational positions or specialties. The PA assumes the duties of the squadron surgeon in his absence. The PA performs general technical health care and administrative duties. The PA works under the clinical supervision of the medical officer. He performs the following duties:

- Establishes and conducts treatment team operations when deployed to other locations away from the squadron aid station.

- Treats, within his ability, sick or injured patients. He refers those patients requiring treatment beyond his capability to the supervising physician.

- Provides EMT and ATM for wounded and DNBI patients.

- Conducts training for squadron personnel in first-aid procedures (self-aid, buddy aid, and CLS), field sanitation, evacuation of the sick and wounded, and the medical aspects of injury prevention.

- Assists in the conduct of the squadron CSC program, to include individual and leader training on the prevention of BF and other stress-related conditions.

- Trains medical personnel in EMT procedures.

D-9 Evacuation Section

The RSTA squadron evacuation section employs interim armored vehicle (IAV) ambulances and provide MEDEVAC through DS or on an area support basis. Each RECON troop will normally have one ambulance team in DS. There are three ambulance teams providing DS and one area support ambulance team that is positioned with the squadron aid station. Each ambulance team consists of an aide/evacuation NCO (E-5) and two ambulance aide/evacuation drivers (E-4 and E-3). Ambulance teams provide MEDEVAC and en route care from either the soldier's point of injury or a CCP to the squadron aid station/ treatment team or brigade medical element providing area support. In mass casualty situations, nonmedical vehicles may be used to assist in casualty evacuation as directed by the commander. Plans for the use of nonmedical vehicles to perform casualty evacuation should be included in the RSTA squadron TSOP and OPORD. The HHT and surveillance and target acquisition troop are provided Echelon I medical treatment and MEDEVAC support on an area support basis by the area support ambulance team. They will also operate dispersed throughout the RSTA squadron AO in support of RECON troops. These dispersed ambulances will evacuate to the nearest supporting Echelon I MTFs based on the OPORD and according to preplanned and coordinated area medical support responsibilities.

D-10. Squadron Combat Health Support Planning

a. The squadron surgeon, assisted by the field medical assistant and the platoon SGT, is responsible for the CHS plan for the RSTA squadron. Squadron CHS operations involve all of the factors that must be considered in the initial developmental stages of the squadron CHS plan. The CHS plan is updated to meet tactical or CHS operations requirements. The following factors should be considered:

- Information requirements (current task organization structure, medical troop strengths, projected weather and environmental factors, and maintenance status of medical equipment).

- Results of the mission analysis.

- Commander's intent.

- Planning guidance.

- Courses of actions.

- Tactical plan.

- Enemy.
- Terrain.
- Troops.
- Weather.
- Threat (including medical threat).
- Operational conditions and constraints.
- Civilian populace in the AO.
- Medical personnel status.
- Equipment status of the medical platoon.
- Supply status including Class VIII.
- Communications capability.
- Nuclear, biological, and chemical defense including OEG.
- Nuclear, biological, and chemical casualty considerations.
- Training status.
- Casualty estimates.
- Medical evacuation requirements and capabilities.
- Nonmedical support requirements from the squadron.
- Area support requirements.
- Special operations support requirement
- Medical records and reports requirements.
- Phases of operations.
- Policy and procedure updates.

b. The foundation of the squadron CHS plan is the squadron commander's guidance and the brigade CHS plan.

D-11. Squadron Operation Plans and Operation Orders

a. The brigade headquarters gives mission orders to the squadron headquarters. The medical platoon may receive additional coordinating instructions from the BSS. These coordinating instructions are normally transmitted in a force text e-mail message via the tactical LAN. As part of the mission analysis and based on the squadron commander's intent and guidance, the medical platoon develops CHS estimates for supporting squadron operations. An understanding of the squadron RECON troop time lines or battle rhythm will assist the squadron medical platoon leader and field medical assistant in developing the CHS input through the squadron S1 to the squadron OPLAN/OPORD.

b. Squadron COA development/analysis and wargaming are accomplished after mission analysis. Course of action development and wargaming result in the production of the OPORD and the CHS annex.

c. Once the RSTA squadron receives the brigade WARNO, it begins mission analysis. Based on its analysis and the full brigade order that follows, the squadron determines its tactical plan. Part of determining its plan is the placement of medical treatment elements (squadron aid station/treatment team). The brigade CHS plan will include the plan for Echelon II support as well as any tasking of support to the squadron medical platoon. The RSTA squadron medical platoon leader will assess the platoon's adequacy as part of the mission analysis. He will bring any shortfalls to the squadron commander's attention during the mission analysis briefing. The platoon leader and/or the field medical assistant then participate in COA development and wargaming to produce the squadron plan. When the commander approves the OPLAN, it becomes the OPORD. The OPLAN and OPORD are developed by the S3 section, using input from each of the staff elements, with the S1 being the staff coordinating element for CHS. The RSTA squadron CHS plan is revised or updated based on mission analysis or changes in CHS requirements. The medical platoon leader maintains current information on the following subject areas that include—

- Patient status board (for example, awaiting evacuation).
- Dirty routes/patient decontamination sites.
- Location of squadron aid station/treatment team and of BAS or other MTF providing area medical support to RSTA squadron elements (current/projected).
- Area medical support responsibilities.

D-12. Rehearsal

For successful implementation of the CHS annex of the RSTA squadron plan, the CHS plan must be coordinated and synchronized with the squadron plan so that CHS requirements are met. To achieve optimal synchronization, the CHS plan is rehearsed as an integral part of the combined arms plan at the combined arms rehearsal. See Chapter 3 for definitive information on rehearsal.

D-13. Combat Health Support for the Offense and the Defense

See Chapter 4 for definitive information on CHS for offensive and defensive operations.

D-14. Reconnaissance, Surveillance, and Target Acquisition Squadron Operations and Combat Health Support

a. The unique mission of the RSTA squadron will cause it to deploy its troops over a very large AO. These long distances will require careful planning and well-coordinated and aggressive CHS operations. This squadron will rely heavily on first aid and CLSs.

b. Many RECON and counter-RECON missions will occur at night. Combat health support for these missions must be planned in detail for limited visibility conditions. The acquisition and MEDEVAC of a RSTA squadron casualty will require a team effort on the part of the troops, the medical platoon, and the squadron staff. Often, trauma specialists will not be able to successfully execute this alone in the RSTA squadron AO. In most cases, it will require a combined arms effort.

D-15. Combat Health Support During Night Operations

The squadron surgeon and medical platoon members must anticipate that the brigade does a substantial amount of its work at night or in limited visibility. They must ensure that the platoon TSOP is available and used throughout the squadron for providing MEDEVAC and treatment at night. See Chapter 4, paragraph 4-4, for definitive information of night operations.

Section III. OVERVIEW OF MEDICAL FORCE STRUCTURE IN SUPPORT OF THE INTERIM BRIGADE

D-16. Combat Health Support for the Interim Brigade

a. Combat health support to the brigade is focused on the stabilization of wounds and injuries, and early evacuation of casualties out of the brigade's AO. Self-aid/buddy aid and the CLS is essential first aid, and early initial medical treatment provided by the trauma specialist is critical for reducing the morbidity and mortality rates. Additionally to reduce the morbidity and mortality rates, both EMT and ATM are performed at the BAS and the BSMC to ensure appropriate treatment and to stabilize the wounded or traumatized patient. After the initial first aid or medical treatment, an ambulance crew evacuates the patient to the supporting Echelon I treatment team/BAS.

b. The BSMC of the brigade support battalion is capable of providing Echelon I area medical support to BSA elements, back-up Echelon I support to forward maneuver battalions, and Echelon II CHS to all elements of the brigade. The BSMC is normally augmented with a surgical capability provided by an attached corps FST. Casualties that cannot be returned to duty by the brigade medical assets will be evacuated to a corps hospital or supporting MTF. After the first 96 hours of brigade operations, the brigade may be augmented with a corps FSMT. When deployed forward to the BSA, the BSMC commander coordinates the air ambulance team's evacuation missions.

c. The BSMC, assisted by the support operations section, provides real-time tactical information to the air ambulance crew about evacuation missions from the brigade combat team units/elements to supporting brigade MTFs. When air ambulances operate forward of the BSA, they will execute the A2C2 plan through the maneuver brigade S3. The BSB support operations section provides planning and coordination between aeromedical evacuation and the supported maneuver brigade. The brigade S3 provides the A2C2 plan that includes the air corridors, air control points, and communications checkpoints. The brigade S3 will provide updates as required. Air ambulances deployed to the BSA provide MEDEVAC from forward areas (BAS) back to the BSA. Air ambulance evacuation from the point of injury will be METT-TC dependent. Corps air ambulances may also evacuate from the BSA to supporting corps MTF. Corps aeromedical elements will operate from BSA providing around the clock immediate response evacuation aircraft. To accomplish this, elements must maintain a close tie with the A2C2 system in the brigade. The brigade A2C2 element provides an airspace plan through the brigade OPORD/OPLAN A2C2 annex. The aircrew must also be familiar with the daily airspace control order and the airspace control plan. These documents contain all ACM, to include free fire areas, no-fly fire areas, restricted operations zones, and established and standard Army aircraft flight routes. These route and ACM change on a daily basis and cannot be integrated into the brigade OPORD. The BSS will ensure all A2C2 information is provided to corps aeromedical elements. The BSS does not generate A2C2 information, but does provide A2C2 planning information to division A2C2 elements.

D-17. Brigade Surgeon's Section

The BSS is assigned to the HHC of the brigade and operates out of the brigade TOC. It is the brigade's primary planning cell for CHS and works closely with the S3 and his staff in the planning process. The BSS staff is responsible to the brigade commander for staff supervision of CHS within the brigade. The BSS is also responsible for coordinating GS and DS relationships of organic medical units and medical units/ elements whether under OPCON or attached to the brigade. The BSS also will—

- Advise the brigade commander on the health of the command.
- Monitor force health protection issues.
- Provide basic sick call support to headquarters personnel.
- Provide technical control/assistance to brigade medical personnel.
- Advise/make recommendations on risk reduction.
- Review/coordinate for either external support or augmentation.

D-18. Brigade Support Medical Company

The BSMC provides Echelon II CHS to all the brigade units operating within the brigade's AO. It also provides/ensures Echelon I CHS to all the brigade units without organic medical support. The BSMC will be organized into a company headquarters, a treatment platoon, an ambulance platoon, a PVNTMED section, and a MH section. The area support section in the treatment platoon contains an area support treatment squad (two treatment teams), which operates the brigade clearing station located in the BSA. These treatment teams will utilize HMMWVs with trailers, or an acceptable variant consistent with the interim brigade concept. The holding squad provides a holding capability of 20 cots. The primary use of the cots is to hold those patients awaiting MEDEVAC out of the brigade AO. An additional mission for the patient holding squad is to hold and provide nursing care for soldiers expected to RTD within 24 to 72 hours. The area support squad provides limited dental support, as well as limited laboratory and x-ray support to the BSMC. The treatment platoon will include two treatment teams to provide mobile Echelon I support within the brigade AO. These treatment teams will utilize IAVs (treatment variant) with trailers. The ambulance platoon will consist of six ambulance teams in IAV ambulances, and four ambulance teams in HMMWV ambulances. The teams in IAV ambulances will be deployed in support of the medical platoons in the maneuver battalions. The IAV ambulances will be used to evacuate patients from BAS to AXP (located between the BAS and the BSMC) where the patients are transferred to BSMC HMMWV ambulances and evacuated to the brigade clearing station operated by the BSMC.

D-19. Forward Surgical Team

Corps-level initial surgical support will be provided by the FST. This team is comprised of 20 personnel and has two operating tables.

D-20. Medical Force Protection

The BSMC medical force protection assets include a PVNTMED section and a MH section. Both sections have two-man teams. The PVNTMED section has an environmental science officer and a PVNTMED specialist. The PVNTMED team provides medical force protection through technical advice to the commander. They ensure the health of the brigade by performing sanitary inspections of food service, field sites, latrine, bathing, and other activities. The team is also responsible for coordination and oversight of medical surveillance. This includes early recognition of potential epidemics or biological warfare agent employment; and monitoring field water supplies, to include sample collection for potential NBC contamination. The MH team includes a behavioral science officer and a MH specialist. The mission of this team is combat operational stress control; they focus on the medical force protection aspects of treatment and prevention of combat stress. This includes the rapid reversal of dysfunctional stress reactions (BF). These medical force protection capabilities are essential to enhancing soldier survivability across the spectrum of battlefield contingencies through continual health hazard assessment, stress reduction, and minimizing disease and injury.

D-21. Medical Logistics (Class VIII/Blood)

The BSMC has limited Class VIII/ blood management capability. During deployment, lodgment, and early buildup phases, medical units operate from planned, prescribed loads and from existing pre-positioned war reserve materiel identified in applicable contingency plans. The BSMC will deploy with supplies to support a 72-hour, self-sustainment mission within the brigade (support for maneuver units is included). Resupply (Class VIII/blood requisitions) for the BSMC will be conducted via a functional module of MC4 electronic requisitions sent to the medical logistics management center, operating out of the AO. The medical logistics management center will send release orders for materiel to the appropriate MEDLOG activity. Class VIII resupply will flow via the battlefield distribution system under the direction of the brigade support battalion support operations section. Unique to the brigade is the range and complexity of MEDLOG support. The probability of corps-level CHS (surgery, hospitalization, intensive care, and extensive blood usage) within the brigade footprint will result in a significant increase in the variety and urgency of medical supplies and equipment. Priority of transportation of critical Class VIII materiel will have to be recognized and supported throughout the distribution pipeline. Blood support to the operation will follow the same procedures.

D-22. Infantry Battalion Medical Platoon

See paragraph D-1.

D-23. Reconnaissance, Surveillance, and Target Acquisition Squadron

See paragraphs D-4 through D-13.

D-24. Field Artillery Battalion

Combat health support for the FA battalion is provided by its medical section that is assigned to the battalion's headquarters service battery. The medical section has a combat medic section with a trauma specialist assigned to support each firing battery. There are three ambulance teams. Normally, one ambulance team is in DS of each firing battery and a treatment team. Echelon II care is provided by the BSMC.

D-25. Engineer Company

Combat health support for the engineer company is provided by its combat medic section. This section is under the supervision of a health care SGT and each platoon is supported by a trauma specialist. Echelon I medical treatment and MEDEVAC support will be provided on an area support basis from brigade medical assets. Echelon II care is provided by the BSMC.

D-26. Antitank Company

Combat health support for the antitank company is provided by its organic medical section that consists of a senior company medic and an individual trauma specialist for each platoon. The antitank platoons will be employed in the proximity of other brigade units that have organic medical treatment teams. Company personnel will receive additional Echelon I from brigade organic medical treatment teams. Echelon II CHS is provided by the BSMC of the brigade support battalion on an area support basis.

D-27. Units Without Organic Medical Personnel

Units without organic medical personnel will utilize self-aid/buddy aid and the CLS according to current directives. These units will normally operate in proximity to area medical support assets. Based on the concepts of employment for these units, personnel will receive Echelons I and II CHS on an area support basis from organic brigade medical assets.

APPENDIX E

FORCE PROTECTION

Section I. UNDERSTANDING FORCE PROTECTION

E-1. Protection

Protection has four components: force protection, field discipline, safety, and fratricide avoidance. Force protection, the primary component, minimizes the effects of enemy firepower (including weapons of mass destruction), maneuver, and information. Field discipline precludes losses from hostile environments. Safety reduces the inherent risk of nonbattle deaths and injuries. Fratricide avoidance minimizes the inadvertent killing or maiming of friendly troops by friendly fires.

E-2 Force Protection

Force protection consists of those actions necessary to prevent or mitigate hostile actions against DOD personnel (to include family members), resources, facilities, and critical information. It coordinates and synchronizes offensive and defensive measures to enable the joint force to perform while degrading opportunities for the enemy. It includes air, space, and missile defense; NBC defense; antiterrorism; defensive information operations; and security to operational forces and means. Force protection does not include actions to defeat the enemy or protect against accidents, weather, and disease. The increased emphasis on force protection at every echelon stems from the conventional dominance of Army forces. Often unable to challenge the Army in conventional combat, opponents seek to frustrate Army operations by resorting to unconventional means, weapons, or tactics. Force protection counters these threats. Force protection at all levels minimizes losses to hostile action. Skillful and aggressive counter-reconnaissance keeps adversaries from seeing the friendly forces while they are defining enemy actions and positions. Effective OPSEC keeps the adversary from exploiting friendly information. Proper dispersion helps reduce losses from enemy fires and terrorist action. Camouflage discipline, security operations, and field fortifications do the same. At the operational level, base protection and lodgment area security contribute to force protection. Theater missile defense forces protect installations and civilian populations from over-the-horizon strikes by conventional warheads and weapons of mass destruction. Army air and missile defense units complement the joint air component's control of the air.

E-3. Field Discipline

Field discipline, a second component of protection, guards soldiers from the physical and psychological effects of the environment. A hostile environment can sap soldier strength and morale far more quickly than enemy action. Soldiers adapt to the point that they out perform their fully acclimatized opponents, who may be native to the region. This mastery can only stem from thorough preparation and training in field craft skills. Commanders take every possible active measure and precaution to keep soldiers healthy and maintain their morale. Such actions include securing equipment and supplies from loss or damage. Commanders ensure systems are in place for adequate CHS, quick return of minor casualties, and PVNTMED. They provide effective systems for maintenance, evacuation, and rapid replacement or repair of equipment. Tactical commanders take care of their soldiers' basic health needs and prevent unnecessary exposure to debilitating conditions.

E-4. Safety

Safety is a third component of protection. Operational conditions often impose significant risks to soldiers' lives and health and make equipment operation difficult. Trained crews and operators must know the capabilities and limitations of their weapon systems. Commanders must know how to employ them. In designing operations, commanders consider the margins of human endurance. They balance the benefits and risks of sustained, high-tempo operations. In combat, fatigue extends reaction times and reduces alertness. Fatal accidents, loss of combat power, and missed tactical opportunities can follow. Command attention to safety and high levels of discipline lessen those risks, particularly as soldiers reach exhaustion. Safe operations come from enforcing standards during training. While taking calculated risks, commanders assume the obligation to embed safety in the conduct of all operations.

E-5. Fratricide Avoidance

A fourth component of protection is fratricide avoidance. *Fratricide* is the unintentional killing or wounding of friendly personnel by friendly firepower. The destructive power and range of modern weapons, coupled with the high intensity and rapid tempo of combat, increase the potential for fratricide. Tactical maneuvers, terrain, and weather conditions may increase the danger of fratricide as well. Commanders seek to lower the probability of fratricide without discouraging boldness and audacity. Situational understanding, positive weapons control, control of troop movements, and disciplined operational procedures, coupled with good leadership, can do this. Reducing fratricide increases soldiers' willingness to act boldly, confident that misdirected friendly fires will not kill them.

Section II. RISK MANAGEMENT, SECURITY MEASURES, AND TERRORISM CONSIDERATIONS

E-6. Risk Management

Risk is the chance of injury or death for individuals and damage to or loss of vehicles and equipment. Risks and/or the potential for risks are always present in every combat and training situation the medical platoon faces. Risk management must take place at all levels of the chain of command during each phase of every operation; it is an integral part of all tactical planning. The battalion commander, company commanders, staff, and all soldiers must know how to use risk management, coupled with fratricide reduction measures, to ensure that the mission is executed in the safest possible environment within mission constraints. For additional information on risk management, refer to FM 100-14.

E-7. Preventive and Protective Security Measures

Preventive and protective security measures should be taken by military units and individual soldiers to protect themselves and their ability to accomplish their mission while deployed. These measures include

OPSEC, communications security, and antiterrorism considerations. The commander develops an antiterrorism plan to institute passive defense measures. The commander must constantly evaluate security plans and measures against the terrorist threat in order to effectively identify security requirements. Terrorism can occur within the US or overseas. Commanders must remain vigilant regardless of where the unit is physically located. Unit training should include orientation to the terrorist threat and countermeasures to be taken at both the individual level and the unit level. For an in-depth discussion of CHS for combatting terrorism operations, refer to FM 8-42.

E-8. Terrorism Considerations

Terrorism considerations that commanders and staffs should consider when they address terrorism is the several relevant characteristics of terrorists and their activities. The first consideration is that anyone can be a victim. (Some terrorists still operate under cultural restraints, such as a desire to avoid harming women, but the planner cannot count on that.) Essentially, there are no innocents. Secondly, attacks which may appear to be senseless and random are not. To the perpetrators, their attacks make perfect sense. Acts, such as bombing public places of assembly and shooting into crowded restaurants, heighten public anxiety. This is the terrorists' immediate objective. Third, the terrorists need to publicize their attack. If no one knows about it, it will not produce fear. The need for publicity often drives the target selection; the greater the symbolic value of the target, the more publicity the attack brings to the terrorists and the more fear it generates. Finally, a leader planning for antiterrorism must understand that he cannot protect every possible target all of the time. He must also understand that terrorists will likely shift from more protected targets to less protected ones. This is the key to defensive measures.

a. Medical units/elements have specific protections afforded to them under the provisions of the Geneva Conventions. The CHS commander/leader must understand that these protections probably will not be recognized nor adhered to by terrorist elements. The CHS commander/leader in developing his force security plan should not consider the Geneva Conventions as a protection from attack by terrorist elements.

b. Terrorists rely on surprise and the victim's confusion at the time of the incident. Antiterrorism involves physical security, OPSEC, and the practice of personal protective measures by all personnel. Commanders and staffs must plan their response to terrorist threats and incidents. Combatting terrorism is an aspect of force protection and is the responsibility of commanders at all levels at all times. Properly planned and executed, the Army antiterrorism program will reduce the probability of surprise while discouraging attack by raising the risk to the attackers. For additional information on combatting terrorism, see Appendix O of FM 8-10-6.

APPENDIX F

TACTICAL STANDING OPERATING PROCEDURE

F-1. General

All medical platoon/sections assigned to combat arms, CS, and CSS units must establish a TSOP for conducting CHS operations supporting their units for all operational contingencies. These TSOP should be detailed and cover all aspects of unit CHS operations. This is an example TSOP that can be used by both the division and nondivisional medical platoons and sections to guide them in the development and refinement of their TSOP.

F-2. Sample Tactical Standing Operating Procedure

The sample shown is from a medical platoon assigned to an infantry battalion and it would be included in an annex of the battalion's TSOP. This TSOP supports the battalion for all operational contingencies and is written for wartime and other operations. There is not a standard format for all TSOP; however, it is recommended that the annex follow the format used by its higher headquarters.

DEPARTMENT OF THE ARMY
Headquarters and Headquarters Company
______Battalion, _______ Infantry
_____________ Division
Fort ________, State/APO________, ZIP CODE________

ANNEX O
MEDICAL PLATOON
TACTICAL STANDING OPERATING PROCEDURES

I. PURPOSE

The purpose of this TSOP is to provide guidelines, policies, and procedures. The implementation of this document will enhance the effectiveness of training and provide specific procedures for routine tasks during CHS operations. This TSOP has been prepared to standardize operations and CHS procedures for the battalion medical platoon in time of war and other operations.

A. Scope. The scope of this TSOP addresses the mission, organization, equipment, and medical platoon operations.

B. Applicability. This SOP applies to all personnel assigned to the ________ Infantry Medical Platoon.

C. Accountability. All personnel assigned to the medical platoon as a part of their initial orientation are required to become familiar with and have a working knowledge of this TSOP. Thereafter, all medical platoon personnel in leadership positions will review the TSOP every 90 days and update or recommend changes as required. Personnel not in leadership positions are required to review the TSOP a minimum of every 6 months or as necessary when conducting operations.

GEORGE I. DOE 2LT, MS Medical Platoon Leader	GREG I. ROCK SFC, USA Medical Platoon Sergeant

II. GENERAL

A. The medical platoon operates the BAS and provides Echelon I CHS for the battalion. Medical platoon personnel are under the command leadership of the battalion commander and the HHC commander. The battalion surgeon/medical platoon leader is a member of the battalion staff. He is responsible to the battalion commander for directing battalion CHS operations. The battalion surgeon/medical platoon leader is responsible to the brigade surgeon for technical oversight and supervision of all CHS activities within the battalion.

B. The medical platoon is dependent on the FSMC of the supporting FSB for Echelon II CHS. This includes medical evacuation from the BAS to the FSMC, patient holding, operational and CSC support, Class VIII resupply, PVNTMED support, medical maintenance, x-ray, laboratory, and operational dental care. The medical platoon requests augmentation/reinforcing support from the FSB's medical company.

III. ORGANIZATION AND MISSION

A. Organization. The medical platoon is organized as shown in Figures F-1.

1. Platoon headquarters.

a. The headquarters section, under the direction of the battalion surgeon, provides C3 and resupply for the platoon. The field medical assistant and the platoon SGT man the platoon headquarters. It is normally collocated with the treatment squad to form the BAS. The CP includes the plans and operations functions performed by the field medical assistant.

b. The field medical assistant, an MS Corps officer, is the operations/readiness officer for the platoon. He is the principal assistant to the battalion surgeon for operations, administration, and logistics. The field medical assistant coordinates CHS operations with the battalion S3 and S4 and coordinates patient evacuation with the FSMC. When a physician or PA are not assigned, he performs the duties of medical platoon leader.

c. The platoon SGT assists the platoon leader and supervises the operations of the platoon. He also serves as the ambulance section SGT. This NCO prepares reports; requests general supplies as well as medical supplies; advises on supply economy procedures; and maintains ASL of expendable supplies. He supervises the activities and functions of the ambulance section, to include operator maintenance of ambulances and equipment; OPSEC; and EMT.

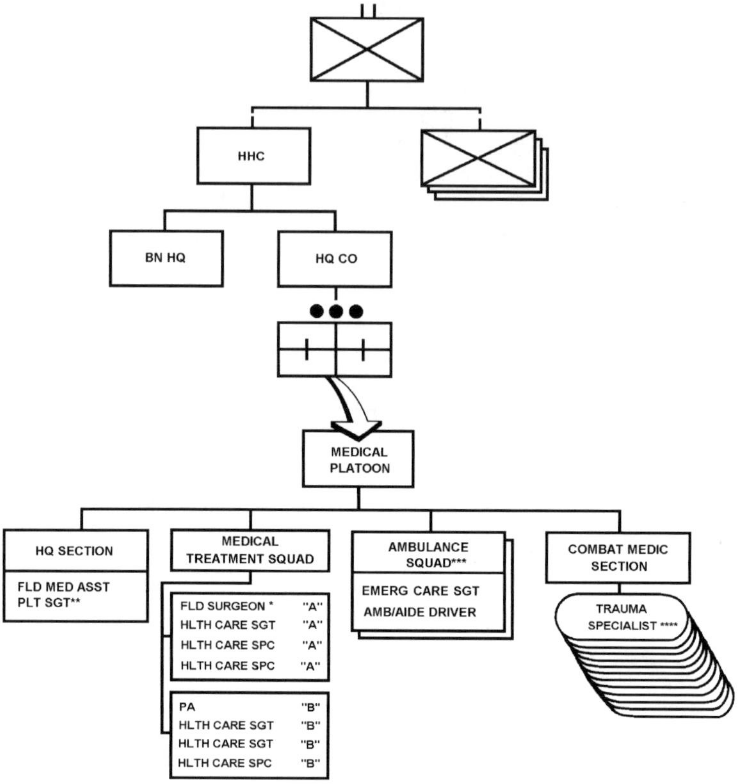

*	ALSO BATTALION SURGEON
**	ALSO SERVES AS AMBULANCE SECTION SGT
***	4 AMBULANCE TEAMS
****	AIRBORNE BATTALIONS HAVE 12 TRAUMA SPECIALISTS ASSIGNED TO THE COMBAT MEDIC SECTION AND AIR ASSAULT BATTALIONS HAVE 10 TRAUMA SPECIALISTS ASSIGNED TO THE COMBAT MEDIC SECTION
"A"	ALFA TREATMENT TEAM
"B"	BRAVO TREATMENT TEAM

Figure F-1. Medical platoon, headquarters and headquarters company light infantry battalion.

2. Treatment squad. This squad is staffed with an operational medical officer (primary care physician/battalion surgeon), a PA, two health care SGT, and four health care specialists. The squad's physician, PA, and health care SGT are all trained in ATM procedures, commensurate with their occupational positions/specialties.

3. Combat medic section. A total of 12 trauma specialists are assigned to the combat medic section. To foster good interpersonal relations and morale of combat troops, every effort should be made to attach the same trauma specialists to the same unit they habitually support each time the unit deploys. However, during lulls in combat operations, they should return to the medical platoon for consultation and proficiency training.

4. Ambulance squads. Medical platoon ambulance squads have two emergency care SGTs, two emergency care specialists, and four ambulance drivers/aides assigned. Four ambulances are assigned to the medical platoon's ambulance squads.

B. Mission. The mission of the medical platoon is to provide Echelon I CHS for the battalion. This includes medical treatment, medical evacuation, and clearing the battlefield. It includes PVNTMED activities to counter either disease or combat and operational stress disorders. It includes EMT and ATM to save lives, limbs, or sight and to stabilize the wounded or injured patient for further evacuation. This also includes maintaining accurate field health records as well as the permanent health record in a garrison setting.

IV. MEDICAL EVACUATION OF SICK AND WOUNDED

A. Purpose. This policy and these procedures established for evacuation of sick and wounded are consistent with doctrine published in the following manuals.

- FM 8-10-1, The Medical Company—Tactics, Techniques, and Procedures.
- FM 8-10-6, Medical Evacuation in a Theater of Operations—Tactics, Techniques, and Procedures.
- FM 8-10-26, Employment of the Medical Company (Air Ambulance).
- FM 8-55, Planning for Health Service Support.

B. General.

1. Evacuation is based on the principle that rear higher echelon medical units are responsible for evacuating patients from supported units. Lower echelon supported and supporting units must ensure evacuation support plans are complete and current by close, direct coordination. See FM 8-10-6 for an in-depth discussion of medical evacuation; for additional information, refer to FMs 4-02.4, 8-10, 8-10-1, 8-10-26, 8-42, 8-55, 63-20, and 63-21.

2. Patients are evacuated no further to the rear than that necessary to obtain the medical care that will return them to duty. Patients are evacuated by the means of transportation that most clearly meets the treatment demands of their wounds, injury, or illness.

3. Allied military personnel, treated or held in a division MTF within reasonable proximity of their own national facility, are classified and processed as follows:

a. Allied military personnel requiring further treatment, but in stable condition for immediate transfer, are returned to their own national medical facility, as coordinated through liaison with the corps or division surgeon.

b. Allied military personnel requiring further stabilization are retained in US medical channels until they can be safely transferred to their own national MTFs. Arrangements for reception of the patient by the gaining MTF are completed prior to the evacuation.

c. The preferred method for evacuation of neuropsychiatric and BF casualties who can be managed without medications or physical restraints is a nonambulance ground vehicle. If physical restraints and/or medications are required during transportation, ground ambulance is preferred. An air ambulance should only be used if no other means of evacuation is available. Physical restraints are used only during transport and medications are given only if needed for reasons of safety. Those neuropsychiatric and BF patients with life- or limb-threatening conditions are evacuated by the most expedient means available. See FMs 8-10-6 and 8-51 for additional information on medical evacuation of psychiatric patient and BF soldiers.

C. Responsibilities for Medical Evacuation.

1. The medical platoon leader—

a. Develops an evacuation plan which will best support the operations being conducted.

b. Prepares/obtains the necessary maps of the AO and overlays from the S3.

c. Does reconnaissance of MEDEVAC routes, either map or on the ground.

d. Provides ambulance teams with strip maps; briefs the plan; and rehearses the MEDEVAC plan with the ambulance section when time permits.

e. Identifies and coordinates with the battalion TOC on the location of primary and alternate helicopter landing sites that are established.

f. Coordinates with the FSMC on medical evacuation and location of the AXP.

g. Oversees medical evacuation operations to ensure expedient evacuation from the battlefield.

2. The medical platoon SGT—

a. Ensures that evacuation wheeled assets are maintained and preventive maintenance checks and services (PMCS) are accomplished in accordance with 10/20 standards.

NOTE

The 10/20 indicate the technical manuals authorized for every item of equipment at unit level.

b. Ensures that ambulances are properly stocked with requisite Class VIII supplies and equipment.

c. Ensures ambulance communication and computer equipment is functional and that all personnel are familiar with SOI.

3. The evacuation squad leader—

a. Participates in route reconnaissance with the medical platoon leader or SGT.

b. Identifies locations of company aid post and CCP.

c. Ensures ambulances are maintained in a state of readiness to perform their evacuation mission.

d. Maintains vehicles in accordance with 10/20 standards and ensures PMCS are performed.

e. Ensures computers and communications equipment are functioning.

f. Keeps the squad updated on road conditions and the threat levels.

4. The senior (line) company trauma specialist—

a. Establishes a company aid post/CCP.

b. Requests medical evacuation of wounded or injured soldiers via the company's communications assets.

c. Coordinates with the BAS as required for additional medical evacuation assets.

d. Maintains aid bags in accordance with published packing lists.

e. Maintains prescribed Class VIII supplies on hand.

f. Request Class VIII resupplies for CLS, as required.

5. Companies (A, B, C, and HHC)—

a. Collect and secure casualty's sensitive items.

b. Remain custodian (1SG) of casualties' personal effects that do not accompany patient through the evacuation process.

c. Ensure the casualty's protective mask accompanies the soldier when he is evacuated from the battalion.

d. Ensure unit personnel are trained on manual carries and on the use of litter carries to transport casualties. Units must be prepared to assist medical personnel as required.

e. Complete the DA Form 1156, Casualty Feeder Report, and DA Form 1155, Witness Statement on Individual, when required. (Medical personnel do not complete these forms.)

6. The battalion S4—

a. Is involved in developing the mass casualty plan and the use of nonstandard vehicles to evacuated casualties.

b. Is responsible for coordinating with graves registration personnel for the transport of deceased personnel (BSA).

c. Provides pouches for human remains (body bags).

d. Provides transportation assets for deceased personnel.

D. Procedures for Medical Evacuation.

1. One ground ambulance, M997, is normally placed in direct support of each company.

2. One ground ambulance, M997, remains with the aid station for supporting the HHC and area support.

3. The two cargo trucks, M998s, used to haul BAS equipment and supplies, may be used for medical evacuation with the augmentation of medical personnel to ride with the patients and provide en route care.

4. Medical platoon ambulances are responsible for medical evacuation of patients from forward areas back to the BAS.

5. The FSMC positions one ambulance, M997, forward with the BAS that is responsible for evacuating patients back to an AXP or to the FSMC in the BSA.

6. One technique used to reduce evacuation times is to pre-position ambulances with the supported company.

7. Another technique is to have on call ambulance support. For this technique, units submit medical evacuation request via radio or landline. The BAS will dispatch an ambulance, M997, to the company aid post/CCP to retrieve the casualty. In some cases, the senior company trauma specialist and the 1SG, using the 1SG's vehicle, may evacuate the casualty from forward areas back to the CCP.

a. Procedures for information collection and MEDEVAC request preparation are performed in accordance with the 9-line MEDEVAC request identified in FM 8-10-6. Air ambulances may not fly forward of the BAS because of threat antiaircraft capabilities. This mission will be performed by a ground ambulance and the following information is collected, to include—

(1) What is the location of the pickup site?

(2) Radio frequency and call sign.

(3) Number of patients by precedence (URGENT, URGENT-SURGICAL, PRIORITY, ROUTINE, and CONVENIENCE).

(4) Special equipment required.

(5) Number of patients by type (litter or ambulatory).

(6) Number and types of wounds (peacetime).

(7) Security of pickup point (wartime).

(8) Method of marking pick-up site if required.

(9) Patient nationality and status.

(10) Nuclear, biological and/or chemical contamination (wartime) or terrain description (peacetime).

b. Upon receipt of the mission, the ambulance team on call will—

- Obtain road clearance from the S3 for the MSR/evacuation route to the requesting unit.

- Perform a radio check with the BAS and/or field trains upon departure from BAS, upon arrival at pickup site, and upon departure from pickup site. Convey problems encountered en route and difficulties regarding en route patient care will also be forwarded.

c. Actions at the unit CCP include—

- Securing (if not already done) site with driver and other available personnel.

- Parking vehicle facing the perimeter (doors to the rear).

- Surveying patients and performing triage (this is accomplished by the senior company trauma specialist on location). Assigning evacuation categories (for example, URGENT, URGENT SURGICAL, PRIORITY, ROUTINE, and TACTICAL CONVENIENCE). Separating NBC-contaminated casualties. Practicing appropriate protective measures in an NBC environment.

- Organizing litter teams (senior company trauma specialist).

- Checking each DD Form 1380 (FMC) to ensure it is completed according to AR 40-66 (senior company trauma specialist).

- Personnel at the CCP assisting with loading patients as directed by the ambulance driver in accordance with Chapter 10, FM 8-10-6.

d. Actions by the ambulance team upon departure of the CCP include—

- Notifying the BAS and/or the field trains as to the time of departure and ETA at the BAS. Stating number and type of patients, to include ambulatory and litter. Discussing potential en route medical complications with physician or PA.

- Moving along prescribed movement routes, taking advantage of cover and concealment. Verifying evacuation routes as tactical situation dictates.

e. Actions of the BAS upon arrival of patients.

- Patients are off loaded and provided EMT/ATM as required; the physician or PA will determine if the patient requires further evacuation. Patient information is entered into the treatment and disposition log and also provided to the battalion S1. If the patient is an URGENT or URGENT-SURGICAL precedence, an air ambulance is requested.

- If a ground ambulance is to be used, the ambulance team from the FSMC is notified of their mission. Normally to minimize the time that the supporting FSMC ambulance is away from the BAS, the FSMC will establish an AXP somewhere between the supported unit and the BSA. This AXP is where the FSMC ambulance team that is supporting the BAS evacuates the patient(s). At the AXP, the patient is transferred to another ground ambulance then evacuated the remaining distance to the FSMC. The medical platoon leader will inform the battalion staff and the patient's company commander on the disposition of his unit member.

- From the BSA, corps ground or air ambulances are responsible for evacuating to the corps combat support hospital.

- Aeromedical evacuations are conducted in accordance with FMs 8-10-6 and 8-10-26.

E. Control of Property and Equipment.

1. Soldiers evacuated from their unit to the BAS, as a minimum, have their protective mask and clothing.

2. Any property and equipment arriving with casualties other than the protective mask and clothing or individual weapon for ambulatory patients will be collected and turned in to the battalion S4 for return to the parent unit. The battalion S4 coordinates the return of property and equipment to the casualty's unit.

3. Under combat conditions, protective masks are kept in the immediate proximity of each patient throughout their period of evacuation and stay at any MTF. In other operations, the protective mask policy for patients will be based on the NBC threat and the policy established by higher headquarters.

F. Rules for Employment of Ambulance and Ambulance Personnel.

1. The use of MEDEVAC vehicles will be restricted to—

a. Transportation of sick or injured personnel.

b. Transportation of medical personnel.

c. Transportation of Class VIII supplies/equipment and blood.

2. Medical personnel assigned to the ambulances will—

a. Adhere to the tactical commanders' standards for uniform and camouflage and other requirements identified in the supported unit's TSOP.

b. Participate in the medical training being conducted at the supported medical element.

c. Assist with patient treatment as required.

NOTE

Caution should be exercised by officers-in-charge or noncommissioned officers-in-charge to ensure the ambulance crew has adequate rest in order that they can safely perform their evacuation duties.

d. Perform PMCS on their vehicles.

e. Ensure their vehicle is restocked with required Class VIII supplies and equipment, full of fuel, and ready for the next evacuation mission.

3. Medical personnel assigned to the ambulances that are positioned with the supported medical element will not be required to—

- Perform duties as kitchen police.
- Perform EPW or perimeter guards.
- Perform driver duties for other than their assigned vehicle.

G. Use of Aeromedical Evacuation.

1. Aeromedical evacuation is the preferred method of evacuation and will be used when—

a. Life, limb, or eyesight is in jeopardy (URGENT or URGENT-SURGICAL category).

b. Speed, distance, and time are factors in assuring prompt and adequate treatment.

c. There is a critical need for resupply of Class VIII supplies or whole blood/blood products.

d. There is a critical need for movement of medical personnel and equipment.

2. Helicopter landing zones are established when and where tactical situations permit. A helicopter landing zone should be marked with a letter "H" or a letter "Y," using identification panels or other appropriate marking material. See FMs 8-10-6 and 57-38 for a complete description and guidelines for establishing a helicopter landing zone.

3. Precedence for air ambulance evacuation is provided in FMs 8-10-6 and 8-10-26.

V. DECEASED PERSONNEL

A. Principles Governing Medical Disposition of Deceased Personnel.

1. Deceased personnel are segregated from other casualties.

2. The deceased, as determined by the senior medical authority, are not evacuated with other casualties, nor are they routinely evacuated on medical vehicles. This is especially true if the threat of biological or chemical contamination will render the vehicle unfit for subsequent medical evacuation missions. A FMC should be initiated and attached to the remains, if possible.

3. Medical evacuation resources should not be used to transport deceased personnel.

4. All deceased personnel should have an FMC that is signed by a medical officer prior to their departure from a graves registration collection point operating in forward areas.

B. Coordinate with the Battalion S4 for Transport of Deceased Personnel.

1. The battalion S4 coordinates with mortuary affairs collection point personnel for the evacuation of deceased personnel to the graves registration site (normally near or in the BSA).

2. The battalion S4 obtains and issues pouches for human remains (body bags) to the BAS.

3. Battalion aid station personnel coordinate transport of deceased personnel with the battalion S4.

4. For definitive information on mortuary affairs operations, see FM 10-64.

VI. ENEMY PRISONERS OF WAR

A. All EPW will be provided medical care according to the articles of the GWS, dated 12 August 1949.

B. Enemy prisoner of war patients will be segregated from allied and US personnel.

C. Enemy prisoner of war patients will be reported through normal medical reporting procedures.

D. Enemy medical personnel are considered retained personnel and shall receive the benefits provided by the Geneva Conventions. Retained enemy medical personnel will be used to the maximum extent possible to care and treat EPW patients.

E. Enemy prisoner of war patients will be evacuated through medical channels.

F. Enemy prisoner of war patients will be under armed guard at all times. Guards are the responsibility of the echelon commander. Medical personnel will not be used as guards for EPW according to the Geneva Conventions.

G. Enemy prisoner of war patients will be searched prior to each move in the MEDEVAC system.

H. Information on EPW patients will be coordinated with the prisoner of war information center to maintain accountability of captives in medical channels. See FM 19-4 for additional information on EPW.

VII. CLASS VIII SUPPLY

A. Class VIII Supply Procedures in the Field. In the field, the medical platoon maintains a 2-day (48-hour) stockage of Class VIII supplies within its MES. The following MES are authorized for the medical platoon treatment section:

- Chemical Agent Patient Decontamination, NSN 6545-01-176-4612 (1).
- Chemical Agent Patient Treatment, NSN 6545-01-141-9469 (2).
- Sick Call Field, NSN 6545-01-228-1886 *(2).
- Trauma Field, NSN 6545-01-228-1667 *(2).

* Indicates the numbers of MES authorized for each treatment team.

B. Authorized Stockage and Controls. Medical supply items authorized for use by the medical platoon are normally those items that are identified as part of the MES. Items that are not in the MES must be approved for stockage by the division surgeon. This includes both expendable items and pharmaceuticals. For perishable and dated items that are found in the MES, a DA Form 4998-R, Quality Control and Surveillance Record for TOE Medical Assemblage, is initiated for all expendable and durable items in the medical assemblage. For a medical item that has a shelf life and demands are expected, a DA Form 4996-R will be prepared and maintained in accordance with AR 40-61. These items are referred to as potency and dated items. Controlled pharmaceutical such as R and Q items (see NOTE below) are stored in the most secure container. A record of controlled medical items will be kept on DA Form 3862 (Controlled Substances Stock Record).

NOTE

Controlled substances are drugs so designated by the Drug Enforcement Administration. A list of these drugs and changes are published in the Federal Register and in the SB 8-75 series. Standard controlled substances are identified by Note R and Q in the notes column of the Federal Supply Catalog, DOD Section, Medical Materiel and by controlled inventory items codes R and Q in the Army Master Data File.

QUALITY CONTROL AND SURVEILLANCE RECORD FOR TOE MEDICAL ASSEMBLAGES
For use of this form, see AR 40-61; the proponent agency is OTSG

NO	LOCATION	MANUFACTURER	CONTRACT NO. (If available)	LOT/BATCH NUMBER	EXP/MFR DATE (If available)	QTY ON HAND	DATE LAST INSPECTION	DATE NEXT INSPECTION

NSN	DESCRIPTION	UNIT OF ISSUE	NOTES	INSPECTION FREQUENCY	SHELF LIFE/ ESTIMATED SHELF LIFE

SAMPLE

DA FORM 4998-R, APR 1994 — REPLACES DA FORM 4998-R, AUG 1981, WHICH IS OBSOLETE — USAPA V1.00

Figure F-2. Quality Control and Surveillance Record for TOE Medical Assemblage, DA Form 4998-R.

NSN	DESCRIPTION	INSP FREQ	DATE LAST INSP	DATE NEXT INSP

QUALITY CONTROL CARD
For use of this form, see AR 40-61; the proponent agency is OTSG

NO	MANUFACTURER	LOT NUMBER	EXP DATE	DATE MFG	SHELF LIFE	DATE RECD
1						
2						
3						
4						
5						
6						
7						
8						
9						
10						

NSN	DESCRIPTION	INSP FREQ	DATE LAST INSP	DATE NEXT INSP

SAMPLE

DA FORM 4996-R, APR 1994 — REPLACES DA FORM 4996-R, AUG 1981, WHICH IS OBSOLETE — USAPA V1.00

Figure F-3. Quality Control Card, DA Form 4996-R.

C. Class VIII Requisitioning Procedures in the Field.

1. Routine Requisitions. Routine requisitions for Class VIII supplies are submitted to the FSMC using a DA Form 2765-1, Request for Issue or Turn-in, for each item in accordance with DA Pam 710-2-1. The FSMC may fill these requests from its MESs or forward them to the DMSO. The DMSO will fill the requests and send items forward to the supporting FSMC via LOGPAC. The FSMC coordinates the delivery of Class VIII supplies to the requesting unit through the FSB support operations section. Requisitions may also be sent via radio or by using a sneaker net method. Prior to deployment, individuals should know the Class VIII resupply requirement and procedures.

2. Emergency Requisitions. Emergency requisitions for Class VIII supplies are sent to the FSMC via radio or telephone. Any item not filled is immediately forwarded to the DMSO. Delivery of emergency requests will be by the most expedient mode of transportation, based on METT-TC factors. Normally, a requisition with a 03 or higher priority will require the approval of the commander.

D. Class VIII Supply Procedures in Garrison.

1. Medical Treatment Facility Operations. Operations of the BAS/MTF at the home station are normally under the control of the medical center (MEDCEN)/MEDDAC commander. For MTF (garrison) operations, Class VIII items may be requisitioned directly from the MEDDAC. Under other situations, the BAS/MTF orders from the DMSO or supporting MEDLOG battalion. The installation medical supply activity is operated by either the MEDDAC or the MEDLOG battalion.

2. Class VIII Requirements. The MEDCEN/MEDDAC commander establishes requirements for MTF operations in garrison. A list of Class VIII items required and authorized at the MTF is identified by hospital SOP or regulation. Medications are identified in the hospital formulary. When operating MTF in a garrison environment, medical platoons are subject to the same standards as any of the MEDCEN/MEDDAC treatment facilities.

3. Requisitioning Information. Requisitioning information includes—

a. Department of Defense activity address code. Example: YNAMTF

b. Serial numbers. The serial number is issued by the document register maintainer and this number is preceded by the Julian date on each request for issue or turn-in. Serial number example: W35MW2.

c. Account processing code. This code is issued by the supporting MEDLOG element and identifies the unit. An example is AWHA.

d. Signature cards will be on file with the MEDCEN/MEDDAC and a copy maintained by the platoon SGT or his designated person for supply actions.

e. The platoon SGT or his designated person for supply activities maintains the document register.

4. Supply Matrix. A supply matrix will be posted in the supply 3-ring binder with the following information:

a. Quantity maintained.

b. Quantity needed to be ordered.

c. Date medication was ordered.

d. Expiration dates of pharmaceuticals (checked each month).

5. Expired Medications. If from the MTF, they are turned in to the MEDCEN/MEDDAC. If from the TOE MES, they are turned in to the DMSO.

6. Multidose Vials. When opened, they are dated and initialed by the user.

7. Medication Refrigerator. Will be cleaned and defrosted monthly.

8. Medication Locker and Safe. Medication lockers are unlocked during sick call but are locked if the care provider leaves the room. The safe with code R and Q items is secured except for removal of medication to fill a prescription. Keys will be maintained by the physician, PA, or platoon SGT. Only the physician, PA, or platoon SGT will have the combination to the safe.

9. Inventories. A 100 percent inventory of all assets belonging to the medical platoon is performed on an annual basis according to DA Pam 710-2-1 and DA Pam 710-2-2.

10. Point of Contact. The telephone number of the MEDCEN/MEDDAC logistics division's is ________________.

VIII. MEDICAL MAINTENANCE

A. Garrison Medical Treatment Facility. Operator maintenance is performed on medical equipment according to MEDCEN/MEDDAC SOP and in accordance with manufacture's instructions. Medical equipment is turned in to the MEDCEN/MEDDAC medical maintenance shop or a work order is phoned in according to MEDCEN/MEDDAC SOP.

B. Medical Maintenance in the Field. Medical maintenance will consist of operator-/user-level maintenance. See FM 4-02.1 for additional information. Medical platoon personnel will exercise their responsibilities by performing operator PMCS. This includes maintaining equipment by performing routine services like cleaning, dusting, washing, and checking for frayed cables, loose hardware, and cracked or rotting seals. In addition, medical platoon personnel will—

- Perform equipment operational testing.

- Replace operator-level spares and repair parts that will not require extensive disassembly of the end item, critical adjustment after the replacement, or extensive use of tools.

IX. MANAGEMENT OF MASS CASUALTIES

A. Mass casualty situations occur when the number of casualties exceed the available medical capability to rapidly treat and evacuate them. The battalion surgeon working with the S4 and the S3 advises the commander on integrating all available resources into an effective mass casualty plan.

B. All companies must have procedures in place to respond effectively to mass casualty situations. The potential of disasters in war and other operations requires that the medical element be prepared to support mass casualty situations. They must be able to receive, triage, treat, and evacuate large numbers of casualties within a short period of time. Contingency plans for supporting mass casualty operations must be developed by all units in coordination with their battalion surgeon and battalion S3. Unit mass casualty plans, as a minimum, will address the following subject areas:

1. Planning and training requirements.

2. Medical duty positions.

3. Nonmedical personnel positions and duties, including litter teams, perimeter guards crowd control, and information personnel.

4. Location of treatment areas, to include triage, immediate care, minimal care, delayed care, and expectant care areas.

5. Support requirements beyond the unit's capability.

6. Medical evacuation.

7. Use of nonmedical transportation assets.

8. Nuclear, biological, and chemical casualties.

9. Return to duty procedures.

10. Medical records and reports.

11. Locating deceased personnel away from and out of sight of all patients.

C. Upon notification of large number of casualties, the medical platoon will request augmentation support from the FSMC. The division or brigade surgeons should be informed of any mass casualty situation by the most expedient means available. As a minimum, information provided should include location, anticipated number of casualties, and additional support required. For additional information, see FMs 8-10, 8-10-1, 8-55.

D. The DSS directs and coordinates CHS requirements for the requesting unit. Supporting corps and division medical units in the chain of evacuation are alerted of the situation.

E. Mass casualty management in garrison is accomplished according to the MEDCEN/MEDDAC SOP.

X. PREVENTIVE MEDICINE

A. The battalion surgeon oversees all PVNTMED activities in the battalion.

B. The division PVNTMED cell located in the MSB (or for Force XXI, the DSB) is responsible for supervising the division's PVNTMED program as described in AR 40-5. This cell ensures PVNTMED measures are implemented to protect division personnel against food-, water-, and arthropodborne diseases, as well as environmental injuries (for example, heat and cold injuries). This cell provides advice and consultation in the areas of environmental sanitation, epidemiology, sanitary engineering, and pest management.

C. Preventive medicine support for the battalion is requested through the FSMC and formal tasking is accomplished through the division or brigade headquarters.

D. All companies in the battalion will establish unit field sanitation teams. Preventive medicine personnel will assist in the training of these teams in the aspects of environmental sanitation and the limited control of animal reservoirs and disease vectors.

E. Company commanders will—

1. Use trained field sanitation team members to assist in preserving the health of the unit and reducing the incidence of DNBI which will hinder mission accomplishment (FM 21-10).

2. Ensure the field sanitation team members take to the field all required field sanitation equipment and supplies to perform their duty (AR 40-5).

3. Enforce food and water safety standards. Unless otherwise stated, water will be treated to at least 5 parts per million chloride residual and will be obtained from approved sources only. Safe handling, storage, and preparation of food will be according to AR 30-21, AR 40-5, FMs 10-23, and 21-10.

4. Plan for the construction of hygienic devices, such as handwashing devices in the unit area. Company commanders will also enforce personal hygiene measures to reduce the threat of disease.

5. Motivate subordinates to execute individual preventive measures (such as carrying an extra pair of dry socks; and/or eating or drinking from approved sources only).

6. Enforce the use of the DOD repellent systems requiring the use of repellents on skin (DEET [75 percent N,N-diethyl-M-tolumide]) and clothing (permethrin).

7. Develop and enforce the unit sleep plan that provides soldiers with a minimum of 4 hours of uninterrupted sleep in a 24-hour period. If sleep is interrupted, then 5 hours should be given. During continuous operations when uninterrupted sleep is not possible, blocks of sleep which add up to 6 hours in a 24-hour period are adequate for most people. Remember, 4 hours each 24-hour period is far from ideal. Do not go with only 4 hours sleep each 24 hours for more than 2 weeks before paying back sleep debt. Recovery time should be approximately 8 to 10 hours sleep each 24 hours over a 5- to 7-day period. If at all possible, give the individuals (such as ambulance drivers) whose key tasks are vulnerable to sleep loss 6 hours of sleep a day.

8. Plan for measures to prevent environmental injuries (such as heat or cold) (see FM 21-10).

9. Obtain and disseminate information on the medical threat so soldiers can reduce their risk of DNBI.

10. Request PVNTMED consultation/assistance. Requests can be submitted to the DSS, the BSS, or any medical company/element in the division.

XI. DENTAL SERVICES

A. Dental treatment facilities are located in each FSMC and in the MSMC (or DSMC for Force XXI). Each medical company establishes dental sick call hours; supported units are notified of the sick call hours.

B. Operational dental care will be provided in the field. This dental care consists of emergency and essential care.

C. Battalion personnel report to the BAS for dental sick call. Dental patients are sent back to the FSMC for dental services.

D. The goal by battalion personnel for dental patients is to relieve pain, to get them back for dental treatment, to get them seen, and to return them to duty.

XII. MENTAL HEALTH/COMBAT STRESS CONTROL

A. Under the guidance of the division psychiatrist and the FSMC MH section, the battalion surgeon is responsible for the battalion's CSC programs.

1. Conduct or request classes on stress control.

2. Identify and eliminate unnecessary stress on unit personnel.

3. Ensure personnel and leaders are familiar with stress-reducing techniques.

4. Ensure all leaders understand the problems associated with sleep deprivation and the consequences of not following the unit sleep plan.

B. The battalion surgeon requests after-action stress debriefings for battalion personnel involved in operations where numerous injuries or loss of life has occurred. He coordinates for critical-events debriefing following a catastrophic incident in the unit. See FMs 8-51 and 22-51 for definitive guidance on stress control.

XIII. EYEWEAR

A. Battalion personnel requiring corrective eyewear will deploy with two pairs of eyewear plus one set of inserts for their protective mask.

B. Personnel authorized to wear contact lenses will deploy with two pair of standard eyewear.

C. The following procedures are followed for repair or replacement of eyewear including—

1. Maintaining the latest prescription on the soldier's field medical record.

2. Sending eyewear that is broken or in need of repair to the division optometry section located in the MSMC (or DSMC for Force XXI).

3. Requesting the replacement of lost eyewear.

XIV. GENEVA CONVENTIONS COMPLIANCE

A. Medical Facilities.

1. All US medical facilities and units, except veterinary, will display the distinctive flag of the Geneva Conventions. This flag consists of a red cross on a white background. It is displayed over the unit or facility and in other places as necessary to adequately identify the unit or facility. Nondisplay of the flag can be ordered by a brigade or higher level commander.

2. Camouflage of the medical facility (medical units, medical vehicle, and medical aircraft on the ground) is authorized when a lack of camouflage might compromise the tactical operation.

3. The order to camouflage can be given by a brigade-level or higher commander.

NOTE

As used in this context, camouflage means to cover up or remove the Geneva Conventions emblem. The black cross on an olive background is not a recognized emblem of the Geneva Conventions.

B. Defense of Medical Units.

1. Medical personnel may carry small arms for personal defense of themselves and defense of their patients. Self-defense of medical personnel or defense by medical personnel of their patients is always permitted. This does not mean that they may resist capture or otherwise fire on the advancing enemy. It means that, if civilian or enemy military personnel are attacking and ignoring the marked medical status of medical personnel, medical transportation, or the medical unit, the medical personnel may provide self-protection. If an enemy military force merely seeks to assume control of a military medical facility or a vehicle for the purpose of inspection and without firing on it, the facility or vehicle may not resist.

2. An overall defense plan may not require medical units to take offensive or defensive actions against enemy troop at any time. If a medical force is part of a defensive area containing nonmedical units, medical personnel may not be responsible for manning part of the overall perimeter. If located in isolation, the medical unit may provide its own local and internal security if other support is not available. However, all soldiers (medical and nonmedical) providing this internal and local security must comply with the requirements in subparagraph 1 above.

XV. MEDICAL REPORTING

Medical reporting requirements for the battalion should be listed in the battalion TSOP. Medical reporting requirements are established by the division and brigade surgeons. Listed below are documents used for reporting or from which information is obtained for reporting purposes.

A. Field Medical Card. A FMC will be initiated for each new patient and for cases required to be carded-for-record only. This will be accomplished according to AR 40-66 and FM 8-10-6. Field Medical Cards will be conspicuously attached to the patient's clothing.

B. Daily Disposition Log. The Daily Disposition Log is maintained by all Echelon I and Echelon II MTFs assigned or attached to the division. Information from this log is extracted, when required, and provided to the S1 or the supported unit requesting the information. The log is also the primary source document for information needed in the preparation of the Patient Summary Report and the Patient Evacuation and Mortality Report.

C. Medical Reports Format. Medical reporting will be accomplished using the FBCB2, FAX, or voice, transmitted via radio/MSE. A manual backup system will be developed. Formats for medical reports are required to maintain consistency and continuity in reporting procedures for information submitted to the BSS and the DSS. Data contained in these reports are required to support the DSS's capability projections and to assist the BSS, HSSO, and FSMC commander in coordinating and planning CHS operations. Data is also extracted for consolidated reporting to higher headquarters. The guidelines presented below should be followed exactly.

1. Each line of information is divided into a number of fields. Each field has a minimum number of alphanumeric characters as indicated in the sample format provided.

2. Each field is separated by a single slash(/).

3. The end of each set of fields is indicated by a double Slash (//).

4. If information from a prior report has not changed, "NC" will be entered in that field (/ NC//).

5. Reports are formatted according to special instructions and reports format.

D. Medical Situational Report, Battalion Aid Station. The MEDSITREP, BAS is a daily patient summary report. This report is used to inform the commander of the battalion's patient, Class VIII, and medical equipment status. This report is submitted daily, covering the events in a 24-hour time period based on time lines provided by the higher headquarters. It is submitted to the supporting medical company. The battalion surgeon (platoon leader) or platoon SGT is responsible for this report. This report could be dispatched via courier, FAX, and/or teletype.

E. Medical Situational Report, Medical Companies. The MEDSITREP, medical companies is a daily patient summary report. This report is submitted daily to the DSS according to time lines provided by higher headquarters. The following information will be included in line six of this report:

1. Status of all assigned and attached ambulances, to include—

a. Total number of ambulances.

b. Number of ambulances that are operational.

c. Number of ambulances that are nonoperational.

2. Status of personnel; identify shortages by AOC or MOS.

3. Treatment of any EPW will be entered in this section.

4. Identify all patients seen during the reporting period with a number and provide the following information in the order provided below:

a. Nationality.

b. Name.

c. Rank.

d. Service number.

e. Unit.

f. Date of birth.

g. Diagnosis.

h. Disposition.

i. Date of disposition.

j. Gaining unit.

5. A hard copy of each aid station's MEDSITREP must accompany the submitting medical company's report.

F. Medical Situational Report, Medical Operations. The MEDSITREP, medical operations is a consolidated patient summary report. This report is consolidated by the DSS and pertains to the previous 24 hours. It is submitted from the DSS daily to the division based on time lines established by the division surgeon.

G. Patient Evacuation and Mortality Report. All Echelons I and II MTFs assigned or attached to the division prepare the Patient Evacuation and Mortality Report. The purpose of this report is to provide a status of patients seen by division MTFs. This is a weekly report compiled as of 2400 each Sunday and distributed each Monday to supported units.

H. Patient Summary Report. The Patient Summary Report provides the status of patients seen by division medical companies and includes their subordinate elements (dental, optometry, MH, or attached units). This report is a weekly report compiled as of 2400 each Sunday. It is prepared by all Echelons I and II MTFs operating in the division AO. It is submitted each Monday to the DSS.

I. Blood Report. The Blood Report is a required report for requesting blood support. Echelon II MTFs will request only Group O Positive and Group O Negative liquid red blood cells.

J. Team Movement Report. The Team Movement Report is used to track the status and location of teams (PVNTMED, CSC, veterinary, ambulance, and treatment teams).

XVI. AUTHORIZED ABBREVIATIONS AND REPORT CODES FOR MEDICAL REPORTS

AUTHORIZED ABBREVIATIONS

ARMS AND SERVICES:

AVN	AVIATION	EMPLOYED MEANS:
ABN	AIRBORNE	MORTARS
AD	AIR DEFENSE	POISON
AMINF	ARMED INFANTRY	ROCKETS
AMPB	AMPHIBIOUS	SABOTAGE
ARMD	ARMORED	TUBEARTY
ARTY	ARTILLERY	MISSILES

AT	ANTITANK
ATAGM	ANTITANK GUIDED MISSILE

COMMAND LEVEL:

AG	ADJUTANT GENERAL
ARMY	ARMY
BDE	BRIGADE
BN	BATTALION
CO	COMPANY
CORPS	CORPS
DIV	DIVISION
GP	GROUP
HQ	HEADQUARTERS
PLT	PLATOON
RGT	REGIMENT

NATIONALITY:

BE	BELGIAN
CA	CANADIAN
GE	GERMAN
NL	NETHERLANDS/HOLLAND
UK	BRITISH
US	AMERICAN

CAUSE OF CASUALTY TO BE USED FOR MASS CASUALTY REPORTING.

ACCIDENT:	AIRCRASH
ACCIDENT:	MARITIME
ACCIDENT:	MOTOR VEHICLE
ACCIDENT:	RAILWAY
ACCIDENT:	FIRE
ACCIDENT:	INDUSTRIAL
ACCIDENT:	POISON
ACCIDENT:	NATURAL DISASTERS
ACCIDENT:	OTHER CAUSES
BATTLE:	CONVENTIONAL
BATTLE:	NUCLEAR
BATTLE:	BIOLOGICAL
BATTLE:	CHEMICAL

GLOSSARY

ABBREVIATIONS AND ACRONYMS

A2C2 Army airspace command and control

ABCA American, British, Canadian, and Australian

ACE analysis and control element

ACM airspace control measures

ACR armored cavalry regiment(s)

ACS armored cavalry squadron

ADA air defense artillery

ADC assistant division commander(s)

ADC-M assistant division commander(s) for maneuver

ADC-OT assistant division commander(s) for operations and training

ADC-S assistant division commander(s) for support

admin administration

AIPD Army Institute for Professional Development

alt alternate

AM amplitude modulated

amb ambulance

AMEDD Army Medical Department

AMEDDC&S Army Medical Department Center and School

AO area of operations

AOC area of concentration

AOE Army of Excellence

AR Army regulation

ARPERCEN Army Reserve Personnel Center

AR–PERSCOM Army Reserve–Personnel Command

ASL authorized stockage level

AT antitank

atk attack

ATM advanced trauma management

attn attention

AXP ambulance exchange point

BAS battalion aid station

BCIS Battlefield Combat Identification System

bde brigade

BF battle fatigue

BICC battlefield information coordination center

BMNT beginning of morning nautical twilight

BMO battalion maintenance officer

bn battalion

BSA brigade support area

BSMC brigade support medical company

BSS brigade surgeon's section

BSU blood supply unit

btry battery

BW biological warfare

C2 command and control

C3 command, control, and communications

C4 command, control, communications, and computers

CA civil affairs

CAC command aviation company

CASEVAC casualty evacuation

cbt combat

CCP casualty collection point

CDR commander

CEMR civilian employee medical record(s)

CHL combat health logistics

CHS combat health support

CI counterintelligence

CIP combat identification panel

cl class

clr clearing

CLS combat lifesaver

cmd command

CNR combat net radio

co company

COA course(s) of action

comm communications

CONUS continental United States

COSCOM corps support command

COTS commercial off-the-shelf

CP command post

CRP common relevant picture

CS combat support

CSC combat stress control

CSM command sergeant major

CSS combat service support

CSSCS Combat Service Support Control System

CTCP combat train command post

CW chemical warfare

DA Department of the Army

DASB division aviation support battalion

DD Department of Defense

DEET N, N-diethyl-M-tolumide

DISCOM division support command

disp disposition

div division

DIVARTY division artillery

DMC distribution management center

DMLSS Defense Medical Logistics Standard Support

DMOC division medical operations center

DMSO division medical supply office

DNBI disease and nonbattle injury(ies)

DOA dead on arrival

DOD Department of Defense

DODD Department of Defense Directive

DODI Department of Defense Instruction

DOW died of wounds

DS direct support

DSA division support area

DSB division support battalion

DSMC division support medical company

DSS division surgeon's section

DTF dental treatment facility

DVE driver's vision enhancer

EAC echelons above corps

EAD echelons above division

EENT end of evening nautical twilight

emerg emergency

EMT emergency medical treatment

EPLRS Enhanced Position Location and Reporting System

EPW enemy prisoner(s) of war

ETA estimated time of arrival

evac evacuation

EW electronic warfare

FA field artillery

FARE forward air refueling equipment

FAS forward aid station

FAX facsimile

FBCB2 Force XXI Battle Command Brigade and Below System

1SG first sergeant

fld field

FLE forward logistics element

FLOT forward line of own troops

flt flight

FM field manual; frequency modulated

FMC US Field Medical Card

FRAGO fragmentary order

FSB forward support battalion

FSC forward support company

FSE fire support element

FSMC forward support medical company

FSMT forward support MEDEVAC team

FSO fire support officer

FST forward surgical team

G1 Assistant Chief of Staff (Personnel)

G2 Assistant Chief of Staff (Intelligence)

G3 Assistant Chief of Staff (Operations and Plans)

G4 Assistant Chief of Staff (Logistics)

G5 Assistant Chief of Staff (Civil Affairs)

G6 Assistant Chief of Staff (Signal)

GC Geneva Convention Relative to the Protection of Civilian Persons in Time of War

GCSS-A Global Combat Support System–Army

GOTS government off-the-shelf

GP general purpose

GPS Global Positioning System

GPW Geneva Contention Relative to the Treatment of Prisoners of War

GS general support

GSO general supply office

GWS Geneva Convention for the Amelioration of the Conditions of the Wounded and Sick in Armed Forces in the Field

GWS (Sea) Geneva Convention for the Amelioration of the Condition of Wounded, Sick, and Shipwrecked Members of Armed Forces at Sea

HHB headquarters and headquarters battery

HHC headquarters and headquarters company

HHT headquarters and headquarters troop

hlth health

HMMWV high-mobility multipurpose wheeled vehicle

HN host nation

hosp hospital

hq headquarters

HREC health record

HSC headquarters and support company

HSMO health service materiel officer

HSSO health service support office(r)

IAV interim armored vehicle

IBCT interim brigade combat team

IEW intelligence and electronic warfare

IHFR improved high frequency radio

IPB intelligence preparation of the battlefield

IR infrared

JTF joint task force

L light

LAN local area network

LD line of departure

ldr leader

LOA limit of advance

LOC line of communications

LOGPAC logistics package(s)

LOGSITREP Logistical Situational Report

LRSD long-range surveillance detachment

LTC lieutenant colonel

LZ landing zone

maint maintenance

MAJ major

MAS main aid station

MC Medical Corps

MC4 Medical Communications for Combat Casualty Care

MCO movement control office(r)

MDMP military decision-making process

MDT Medical Detachment–Telemedicine

MED medical

MEDCOM medical command

MEDDAC medical department activity

MEDEVAC medical evacuation

MEDLOG medical logistics

MEDSITREP Medical Situational Report

MES medical equipment set(s)

METT-TC mission, enemy, terrain, troops, time available, and civilian considerations

MH mental health

MI military intelligence

MLRS multiple launch rocket system

MMC materiel management center

MMMB medical materiel management branch

MOPP mission-oriented protective posture

MOS military occupational specialty

MP military police

MRI Medical Reengineering Initiative

MRO medical regulating office(r)

MS Medical Service Corps

MSB main support battalion

MSE mobile subscriber equipment

msg message

MSMC main support medical company

MSR main supply route

MTF medical treatment facility(ies)

MTOE modification table(s) of organization and equipment

NATO North Atlantic Treaty Organization

NBC nuclear, biological, and chemical

NCO noncommissioned officer

NCS net control station

NSN national stock number

NVG night vision goggles

O&I operation and intelligence

OAKOC obstacles; avenues of approach; key terrain; observation and fields of fire; and cover and concealment

OBJ objective

OEG operational exposure guide

off officer

OPCON operational control

OPLAN operation plan

OPORD operation order

opr operator

ops operations

OPSEC operations security

OR operating room

PA physician assistant

pam pamphlet

PIC personal information carrier

PL phase line

plt platoon

PMCS preventive maintenance checks and services

pnt patient

POL petroleum, oils and lubricants

PSYOP psychological operations

PVNTMED preventive medicine

PZ pickup zone

QRF quick reactionary force

QSTAG Quadripartite Standardization Agreement

R&S reconnaissance and surveillance

RAS regimental aviation squadron

RECON reconnaissance

regt regiment

rpts reports

RSTA reconnaissance, surveillance, and target acquisition

RT receiver/transmitter

RTD return to duty

S1 Adjutant (US Army)

S2 Intelligence Officer (US Army)

S3 Operations and Training Officer (US Army)

S4 Logistics Officer (US Army)

S6 Communications Staff Officer (US Army)

S&T supply and transportation

sct scout

SEAD suppression of enemy air defense

sec section

2LT second lieutenant

SF standard form

SFC sergeant first class

SGT sergeant

SINCGARS single-channel ground and airborne radio system

sm small

SOI signal operating instructions

SOP standing operating procedure(s)

SP start point

SPC specialist

spt support

SQD squad/squadron

sta station

STANAG Standardization Agreement

TACP tactical air control party

TAMMIS Theater Army Medical Management Information System

TC track commander

TCF tactical combat force

TCON teleconsultation

TDA table(s) of distribution and allowances

TF task force(s)

TMEM telementoring

TMIP Theater Medical Information Program

TOC tactical operations center

TOE table(s) of organization and equipment

trmt treatment

TSOP tactical standing operating procedure(s)

TTP tactics, techniques, and procedures

UMT unit ministry team

US United States

USA United States Army

USAF United States Air Force

USN United States Navy

veh vehicle

WAN wide area network

WARNO warning order

whl wheeled

WIA wounded in action

WIN Warfighter Information Network

XO executive officer

REFERENCES

SOURCES USED

These are the sources quoted or paraphrased in this publication.

NATO STANAGs

These agreements are available on request using DD Form 1425 from Standardization Document Order Desk, 700 Robin Avenue, Building 4, Section D, Philadelphia, Pennsylvania 19111-5094.

2128. *Medical and Dental Supply Procedures. Edition 4.* 21 November 1991.
2454. *Regulations and Procedures for Road Movement and Identification of Movement Control and Traffic Control Personnel and Agencies. Edition 1.* 6 July 1998.
2931. *Orders for the Camouflage of the Red Cross and Red Crescent on Land in Tactical Operations. Edition 2.* 19 January 1998. (Latest Amendment, 3 April 1998.)

ABCA QSTAGs

This agreement is available on request using DD Form 1425 from the Standardization Document Order Desk, 700 Robins Avenue, Building 4, Section D, Philadelphia, Pennsylvania 19111-5094.

435. *Medical Materiel Management During Patient Evacuation. Amendment 1.* 20 June 1984.
436. *Minimum Labelling Requirements for Medical Materiel. Amendment 1.* 8 October 1980.

Joint and Multiservice Publications

Joint Pub 3-07.3. *Joint Tactics, Techniques, and Procedures for Peace Operations.* 12 February 1999.
Joint Pub 4-02. *Doctrine for Health Service Support in Joint Operations.* 26 April 1995.
AR 600-8-14. *Identification Cards for Members of the Uniformed Services, Their Family Members, and Other Eligible Personnel.* AFI 36-3026(I); BUPERS I 1750.10A, Change 2; MCO P5512.11B Change 2; CG M5512.1; Commissioned Corps Personnel Manual 29.2, Instructions 1 and 2; NOAA Corps Regulations, Chapter 1, Part 4. 1 March 1998. (Change 1, 10 August 1999.)
FM 3-3. *Chemical and Biological Contamination Avoidance.* FMFM 11-17. 16 November 1992. (Change 1, 29 September 1994.)
FM 101-5-1. *Operational Terms and Graphics.* MCRP 5-2A. 30 September 1997.
TM 8-227-12. *Armed Services Blood Program Joint Blood Program Handbook.* NAVMED P-6530; AFH 44-152. 21 January 1998.

Department of Defense Directive (DODD)

6490.2. *Joint Medical Surveillance.* 30 August 1997.

Department of Defense Instructions (DODI)

1400.32. *DOD Civilian Work Force Contingency and Emergency Planning Guidelines and Procedures*. 24 April 1995.
3020.37. *Continuation of Essential DOD Contractor Services During Crises*. 6 November 1990. (Administrative Reissuance Incorporating Change 1, January 26, 1996.)
6490.3. *Implementation and Application of Joint Medical Surveillance for Deployments.* 7 August 1997.

Army Regulations (AR)

40-3. *Medical, Dental, and Veterinary Care.* 30 July 1999.
40-5. *Preventive Medicine*. 15 October 1990.
40-61. *Medical Logistics Policies and Procedures*. 25 January 1995.
40-66. *Medical Record Administration and Health Care Documentation.* 3 May 1999.
40-68. *Quality Assurance Administration.* 20 December 1989.
40-501. *Standards of Medical Fitness*. 30 August 1995. (Reprinted with basic including Change 1, 27 February 1998.)
310-25. *Dictionary of United States Army Terms (Short Title: AD).* 15 October 1983. (Reprinted with basic including Change 1, 21 May 1986.)
350-41. *Training in Units*. 19 March 1993.
351-3. *Professional Education and Training Programs of the Army Medical Department.* 8 February 1988.
750-1. *Army Materiel Maintenance Policy and Retail Maintenance Operations*. 1 August 1994. (Change 1, 1 July 1996.)

Field Manuals (FM)

3-19-1. *Military Police Operations*. 22 March 2001.
4-02.1. *Combat Health Logistics*. Scheduled to be printed FY 2001.
4-02.17 (8-10-17). *Preventive Medicine Services*. 28 August 2000.
4-02.19 (8-10-19). *Dental Service Support in a Theater of Operations*. 1 March 2001.
4-02.21. *Division and Brigade Surgeons' Handbook (Digitized)—Tactics, Techniques, and Procedures*. 15 November 2000.
4-02.24 (8-10-24). *Area Support Medical Battalion—Tactics, Techniques, and Procedures*. 28 August 2000.
8-10-3 (4-02.3). *Division Medical Operations Center—Tactics, Techniques, and Procedures*. 12 November 1996.
8-10-5 (4-02.5). *Brigade and Division Surgeons' Handbook—Tactics, Techniques, and Procedures*. 10 June 1991.
8-10-8 (4-02.8). *Medical Intelligence in a Theater of Operations*. 7 July 1989.
8-55 (4-02.55). *Planning for Health Service Support*. 9 September 1994.
10-64. *Mortuary Affairs Operations*. 16 February 1999.
19-4. *Military Police Battlefield Circulation Control, Area Security, and Enemy Prisoner of War Operations*. 7 May 1993.
21-10-1. *Unit Field Sanitation Team*. 11 October 1989.

24-1. *Signal Support in the AirLand Battle*. 15 October 1990.
27-10. *The Law of Land Warfare*. 18 July 1956. (Reprinted with basic including Change 1, 15 July 1976.)
41-10. *Civil Affairs Operations*. 14 February 2000.
57-38. *Pathfinder Operations*. 9 April 1993.
63-2. *Division Support Command, Armored, Infantry, and Mechanized Infantry Divisions*. 20 May 1991.
63-2-1. *Division Support Command, Light Infantry, Airborne, and Air Assault Divisions*. 16 November 1992. (Reprinted with basic including Change 1, 20 September 1994.)
63-2-2. *Division Support Command (Digitized)*. Scheduled to be printed FY 2001.
63-21. *Main Support Battalion*. 7 August 1990.
63-23-2. *Division Aviation Support Battalion (Digitized)*. Scheduled to be printed FY 2001.
100-5. *Operations*. 14 June 1993.
100-7. *Decisive Force: The Army in Theater Operations*. 31 May 1995.
100-10. *Combat Service Support*. 3 October 1995.
100-14. *Risk Management*. 23 April 1998.

DOCUMENTS NEEDED

These documents must be available to the intended users of this publication.

Joint and Multiservice Publications

FM 4-02.283 (Final Draft). *Treatment of Nuclear and Radiological Casualties*. NTRP 4-02.21; AFMAN 44-161(I); MCRP 4-11.1B. Scheduled to be printed FY 2001.
FM 8-284 (4-02.284). *Treatment of Biological Warfare Agent Casualties*. NAVMED P-5042; AFMAN (I) 44-156; MCRP 4-11.1C. 17 July 2000.
FM 8-285 (4-02.285). *Treatment of Chemical Agent Casualties and Conventional Military Chemical Injuries*. NAVMED P-5041; AFJMAN 44-149; FMFM 11-11. 22 December 1995.
*FM 21-10 (4-25.10). *Field Hygiene and Sanitation*. MCRP 4-11.1D. 21 June 2000.

Field Manuals (FMs)

7-20. *The Infantry Battalion*. 6 April 1992. (Change 1, 29 December 2000.)
7-30. *The Infantry Brigade*. 3 October 1995. (Change 1, 31 October 2000.)
*8-10 (4-02). *Health Service Support in a Theater of Operations*. 1 March 1991.
*8-10-1 (4-02.6). *The Medical Company—Tactics, Techniques, and Procedures*. 29 December 1994.
*8-10-6 (4-02.2). *Medical Evacuation in a Theater of Operations—Tactics, Techniques, and Procedures*. 14 April 2000.
*8-42 (4-02.42). *Combat Health Support in Stability Operations and Support Operations*. 27 October 1997.

*This source was also used to develop this publication.

*8-51 (4-02.51). *Combat Stress Control in a Theater of Operations—Tactics, Techniques, and Procedures.* 29 September 1994. (Change 1, 30 January 1998.)
17-95. *Cavalry Operations.* 24 December 1996.
22-51 (4-02.22). *Leaders' Manual for Combat Stress Control.* 29 September 1994.
*63-20. *Forward Support Battalion.* 26 February 1990.
63-20-1. *Forward Support Battalion (Digitized).* Scheduled to be printed FY 2001.
71-3. *The Armored and Mechanized Infantry Brigade.* 8 January 1996.

Department of the Army Forms (DA Form)

145. *Army Correspondence Course Enrollment Application.* October 2000.
1155. *Witness Statement on Individual.* 1 June 1966.
1156. *Casualty Feeder Report.* 1 June 1966.
2765-1. *Request for Issue or Turn-In.* April 1976.
3444. *Terminal Digit File for Treatment Record (Orange).* May 1991.
3862. *Controlled Substances Stock Record.* 1 June 1972.
4996-R. *Quality Control Card (LRA).* August 1981.
4998-R. *Quality Control and Surveillance Record for TOE Medical Assemblage (LRA).* August 1981.
8005. *Outpatient Medical Record (OMR) Orange.* November 1991.
8007-R. *Individual Medical History (LRA).* November 1996.

Department of Defense Forms (DD Form)

1380. *US Field Medical Card.* December 1991.
1425. *Specifications and Standards Requisition.* March 1986.
1934. *Geneva Conventions Identity Card for Medical and Religious Personnel Who Serve in or Accompany the Armed Forces.* July 1974.
2697. *Report of Medical Assessment.* February 1995.
2766. *Adult Preventive and Chronic Care Flowsheet.* March 1998.
2766C. *Adult Preventive and Chronic Care Flowsheet (Continuation Sheet).* March 1998.
2795. *Pre-Deployment Health Assessment Questionnaire.* May 1999.
2796. *Post-Deployment Health Assessment Questionnaire.* May 1999.

Standard Forms (SF)

558. *Medical Record—Emergency Care and Treatment.* September 1996.
600. *Health Record—Chronological Record of Medical Care.* June 1997.
603. *Health Record—Dental.* October 1975.

READINGS RECOMMENDED

These readings contain relevant supplemental information.

Joint and Multiservice Publications

AR 190-8. *Enemy Prisoners of War, Retained Personnel, Civilian Internees, and Other Detainees.* OPNAVINST 3461.6; AFJI 31-304; MCO 3461.1. 1 October 1997.

FM 3-4. *NBC Protection.* FMFM 11-9. 29 May 1992. (Reprinted with basic including Changes 1—2, 21 February 1996.)

FM 3-5. *NBC Decontamination.* MCWP 3-37.3. 28 July 2000.

FM 3-100. *Chemical Operations Principles and Fundamentals.* MCWP 3-3.7.1. 8 May 1996.

FM 8-9 (4-02.11). *NATO Handbook on the Medical Aspect of NBC Defensive Operations AMEDP-6(B), Part I—Nuclear, Part II—Biological, Part III—Chemical.* NAVMED P-5059; AFJMAN 44-151V1V2V3. 1 February 1996.

FM 90-13. *River Crossing Operations.* MCWP 3-17.1. 26 January 1998.

Army Regulations (AR)

30-21. *The Army Field Feeding System.* 24 September 1990.

71-32. *Force Development and Documentation—Consolidated Policies.* 3 March 1997.

700-138. *Army Logistics Readiness and Sustainability.* 16 September 1997.

Department of the Army Pamphlets (DA Pam)

27-1. *Treaties Governing Land Warfare.* 7 December 1956.

350-59. *Army Correspondence Course Program Catalogue.* 1 October 2000.

710-2-1. *Using Unit Supply System (Manual Procedures).* 31 December 1997.

710-2-2. *Supply Support Activity Supply System (Manual Procedures).* 30 September 1998.

Field Manuals (FM)

1-111. *Aviation Brigades.* 27 October 1997.

3-19.1. *Military Police Operations.* 22 March 2001.

3-50. *Smoke Operations.* 4 December 1990. (Change 1, 11 September 1996.)

5-71-100. *Division Engineer Combat Operations,* 22 April 1993.

5-424. *Theater of Operations Electrical Systems.* 25 June 1997.

6-20-2. *Tactics, Techniques, and Procedures for Corps Artillery, Division Artillery, and Field Artillery Brigade Headquarters.* 7 January 1993.

7-92. *The Infantry Reconnaissance Platoon and Squad (Airborne, Air Assault, Light Infantry).* 23 December 1992.

8-10-7 (4-02.7). *Health Service Support in a Nuclear, Biological, and Chemical Environment.* 22 April 1993. (Reprinted with basic including Change 1, 26 November 1996.)

8-10-9 (4-02.9). *Combat Health Logistics in a Theater of Operations—Tactics, Techniques, and Procedures.* 3 October 1995.

8-10-18 (4-20.18). *Veterinary Services—Tactics, Techniques, and Procedures.* 22 August 1997.

8-10-25 (4-02.25). *Employment of Forward Surgical Team—Tactics, Techniques, and Procedures.* 30 September 1997.
8-10-26 (4-02.26). *Employment of the Medical Company (Air Ambulance).* 16 February 1999.
10-23. *Basic Doctrine for Army Field Feeding and Class I Operations Management.* 18 April 1996.
11-43. *The Signal Leader's Guide.* 12 June 1995.
11-50. *Combat Communications Within the Division (Heavy and Light).* 4 April 1991.
11-55. *Mobile Subscriber Equipment (MSE) Operations.* 22 June 1999.
12-6. *Personnel Doctrine.* 9 September 1994.
16-1. *Religious Support.* 26 May 1995.
17-98. *Scout Platoon.* 10 April 1999.
19-40. *Enemy Prisoners of War, Civilian Internees and Detained Persons.* 27 February 1976.
20-3. *Camouflage Concealment, and Decoys.* 30 August 1999.
24-24. *Signal Data References: Signal Equipment.* 29 December 1994.
34-3. *Intelligence Analysis.* 15 March 1990.
34-54. *Technical Intelligence.* 30 January 1998.
44-100. *U.S. Army Air and Missile Defense Operations.* 15 June 2000.
63-21-1. *Division Support Battalion (Digitized).* Scheduled to be printed FY 2001.
63-23-2. *Division Aviation Support Battalion (Digitized).* Scheduled to be printed FY 2001.
63-3. *Corps Support Command.* 30 September 1993.
71-100 (3-71). *Division Operations.* 28 August 1996.
100-103. *Army Airspace Command and Control in a Combat Zone.* 7 October 1987.
101-5. *Staff Organization and Operations.* 31 May 1997.

Supply Bulletins (SB)

8-75-Series. These bulletins provide Army Medical Department supply information.

Technical Bulletin (TB)

38-750-2. *Maintenance Management Procedures for Medical Equipment.* 12 April 1987. (Reprinted with basic including Changes 1—3, 1 November 1989.)

INDEX

References are to paragraph numbers except where specified otherwise.

***FM 4-02.51 (FM 8-51)**

Field Manual
No. 4-02.51 (8-51)

Headquarters
Department of the Army
Washington, DC, 6 July 2006

COMBAT AND OPERATIONAL STRESS CONTROL

Contents

Distribution Restriction: Approved for public release; distribution is unlimited.

***This publication supersedes FM 8-51, 29 September 1994.**

Figures

Tables

Preface

This publication outlines the functions and operations of each combat and operational stress control (COSC) element within an area of operations (AO). This field manual (FM) establishes Army doctrine and provides guidance for conducting COSC support for combat, stability, and reconstruction operations from brigade to theater level. The information provided in this publication will assist commanders and their staffs to operate efficiently at all levels of command and throughout the operational continuum. It may be used by medical planners to supplement FM 4-02, FM 8-42 (4-02.42), and FM 8-55 (4-02.55). Users of this publication should be familiar with Army Regulation (AR) 40-216 and with FM 4-02, FM 4-02.6, FM 4-02.10, FM 4-02.21, FM 4-02.24, FM 6-22.5, FM 8-10-6, FM 8-10-14, and FM 22-51.

This manual is in consonance with FM 7-15, Army Universal Task List (AUTL) and support the following Army tactical task (ART) provided below. Commanders should use the AUTL as a cross-reference for tactical tasks. The AUTL provides a standard doctrinal foundation and catalogue of the Army's tactical collective tasks.

AUTL ART	
ART 6.5.1	Provide Combat Casualty Care
ART 6.5.1.5	Provide Mental Health/Neuropsychiatric Treatment
ART 6.5.4	Provide Casualty Prevention
ART 6.5.4.5	Provide Combat Operational Stress Control Prevention

The staffing and organization structure presented in this publication reflects those established in the base tables of organization and equipment (base TOE) and are current as of the publication print date. Such staffing is subject to change to comply with manpower requirements criteria outlined in AR 71-32. Those requirements criteria are also subject to change if the modification table of organization and equipment (MTOE) is significantly altered.

Users of this publication are encouraged to submit comments and recommendations to improve the publication. Comments should include the page, paragraph, and line(s) of the text where the change is recommended. The proponent of this publication is the United States (US) Army Medical Department Center and School (USAMEDDC&S). Comments and recommendations should be forwarded in a letter format directly to the **Commander, USAMEDDC&S, ATTN: MCCS-FCD-L, 1400 East Grayson Street, Fort Sam Houston, Texas 78234-5052**, or at e-mail address: Medicaldoctrine@amedd.army.mil.

This FM applies to the Active Army, the Army National Guard (ARNG)/Army National Guard of the United States (ARNGUS), and the United States Army Reserve (USAR) unless otherwise stated.

Unless this publication states otherwise, masculine nouns and pronouns do not refer exclusively to men.

Use of trade or brand names in this publication is for illustrative purposes only and does not imply endorsement by the Department of Defense (DOD). The Army Medical Department (AMEDD) is in a transitional phase with terminology. This manual uses the most current terminology; however, other FM 4-02-series and FM 8-series may use the older terminology. Changes in terminology are a result of adopting the terminology currently used in the joint and/or North Atlantic Treaty Organization (NATO) and American, British, Canadian, and Australian (ABCA) Armies publication arenas. Therefore, the following terms are synonymous and the current terms are listed first, to include—

- Force health protection (FHP), health service support (HSS), and combat health support (CHS).
- Medical logistics (MEDLOG), health service logistics (HSL), and combat health logistics (CHL).
- Levels of care, echelons of care, and roles of care.

- Combat and operational stress control (COSC) and combat stress control (CSC).
- Behavioral health (BH) and mental health (MH).

Additionally, please note that the term "battle fatigue (BF)" that is used in AR 40-216 is being replaced with the term "combat stress reaction (CSR)." This change is due to DOD Directive (DODD) 6490.5, which specified that all Military Services use the term CSR for the purpose of joint interoperability. A DOD (Health Affairs) working group with the Services later added the term "operational stress reaction (OSR)" to further characterize stress reactions experienced by Soldiers. The only difference between a CSR and an OSR is that it takes place in a combat environment. To reduce confusion from this change in terminology, this manual will use combat and operational stress reaction (COSR) as opposed to "BF". The COSR (battle fatigue) "casualties" are Soldiers experiencing a stress reaction in combat or operational environment.

Introduction

In our own Soldiers and in the enemy combatants, control of stress is often the decisive difference between victory and defeat across the operational continuum. Battles and wars are won more by controlling the will to fight than by killing all of the enemy combatants. Uncontrolled combat stress causes erratic or harmful behaviors, impair mission performance, and may result in disaster and defeat of COSC preventive measures. The COSC preventive measures are aimed at minimizing maladaptive stress reactions while promoting adaptive stress reactions, such as loyalty, selflessness, and acts of bravery.

This manual provides doctrinal guidance for controlling excessive stress in combat and other operational environments. It identifies command and leadership responsibilities for COSC. It identifies COSC consultation, training, and education assistance available for units. This manual provides definitive guidance to BH personnel and CSC units for their COSC mission and for management of COSR and other behavioral disordered patients (BDPs). It identifies the requirements for COSC consultation, planning, coordination, rehearsal, and implementation of the COSC plan contained in the FHP annex of the operation order (OPORD).

Many stressors in a combat situation are due to deliberate enemy actions aimed at killing, wounding, or demoralizing our Soldiers and our allies. Other stressors are due to the operational environment. Some of these stressors can be avoided or counteracted by wise command actions. Still other stressors are due to our own calculated or miscalculated choice, accepted in order to exert greater stress on the enemy. Sound leadership works to keep stressors within tolerable limits and prepares the troops mentally and physically to endure them. Some of the most potent stressors can be due to personal organizational problems in the unit or on the home front. These, too, must be identified and, when possible, corrected or controlled. Unit needs assessments (UNAs) can help BH providers identify specific stressor in a unit and develop interventions to help unit personnel cope.

This manual identifies MH sections and medical units, CSC, that provides COSC support to units. It identifies the requirements for COSC assistance, as a unit transitions through the different phases of a deployment. Chaplains, commander, leader and COSC Soldier mentors provide assistance with after-action review (AAR) and traumatic events management (TEM).

Chapter 1

Combat and Operational Stress Control

SECTION I — INTERVENTION AND CONTROL FOR THE COMBAT AND OPERATIONAL STRESS THREAT

STRESS CONTROL

1-1. Combat stress includes all the physiological and emotional stresses encountered as a direct result of the dangers and mission demands of combat, see AR 40-216. Combat and operational stress control in the Army may be defined as programs developed and actions taken by military leadership to prevent, identify, and manage adverse COSRs in units. This program optimizes mission performance; conserves the fighting strength; and prevents or minimizes adverse effects of COSR on Soldiers and their physical, psychological, intellectual, and social health. Its goal is to return Soldiers to duty expeditiously. According to DODD 6490.2, COSC activities include routine screening of individuals when recruited; continued surveillance throughout military service, especially before, during, and after deployment; continual assessment and consultation with medical and other personnel from garrison to the battlefield.

1-2. Combat and operational stress control is the commander's responsibility at all levels. The commander is assisted with his responsibility for COSC by his staff, unit leaders, unit chaplain (Appendix A), and organic medical personnel. The commander may also receive assistance from organic COSC personnel at brigade and above, and from corps and above medical company/detachment CSC BH personnel. The key concern to combat commanders is to maximize the return-to-duty (RTD) rate of Soldiers who are temporarily impaired or incapacitated with stress-related conditions or diagnosed behavioral disorders.

1-3. The purpose of COSC is to promote Soldier and unit readiness by—

- Enhancing adaptive stress reactions.
- Preventing maladaptive stress reactions.
- Assisting Soldiers with controlling COSRs.
- Assisting Soldiers with behavioral disorders.

> ***Note.*** The word control is used with combat and operational stress, rather than the word management, to emphasize the active steps that leaders, supporting BH personnel, and individual Soldiers must take to keep stress within an acceptable range.

COMBAT AND OPERATIONAL STRESS THREAT

1-4. Many stressors in a combat situation are due to deliberate enemy actions aimed at killing, wounding, or demoralizing our Soldiers and our allies. Other stressors are due to the natural environment, such as intense heat or cold, humidity, or poor air quality. Still others are due to leaders' own calculated or miscalculated choices (for example, decisions about unit strength, maneuver, the time of the attack, and plans for medical and logistical support). Sound leadership works to keep operational stressors within tolerable limits and prepares troops mentally and physically to endure them. In some cases however, excessive stress can affect both leaders' and Soldiers' decision-making and judgment, resulting in missed opportunities, or worse, in high casualties and/or failure to complete the mission. Finally, some of the most potent stressors are interpersonal in nature and can be due to conflict in the unit or on the home front. Extreme reactions to such stressors may involve harm to self (as in the hypothetical case of a Soldier that becomes suicidal on discovering that his wife wants a divorce) or to others (as in the case of a Soldier that impulsively fires a weapon at his noncommissioned officer (NCO) out of rage over perceived unfairness). These stressors must be identified and when possible, corrected or controlled. For more information on the control of combat/operational stressors and for details about specific leader actions to control stress, see FM 22-51. Also see FM 4-02 and FM 4-02.17 for additional information on the general and medical threat.

EFFECT OF STRESS

1-5. Focused stress is vital to survival and mission accomplishment. However, stress that is too intense, or prolonged, results in COSR that impairs their ability to function effectively. Some stressors contribute to misconduct that requires disciplinary action and may take the Soldier from duty for legal action and incarceration. In a broader context stress may cause battle and nonbattle injuries through inattention, clumsiness, and reckless behavior. These resultant injuries can include equipment losses and friendly fire incidents. Stress may increase disease rates by disrupting hygiene and protective measures, and impairing the body's immune defenses. Stress may progress to BH disorders, or suicidal behaviors and/or homicidal behaviors. Excessive stress in combat contributes to lapses in operational and tactical judgment and to missed opportunities that could increase the numbers of Soldiers injured over time.

MODEL FOR COMBAT AND OPERATIONAL STRESS CONTROL INTERVENTIONS

1-6. Combat and operational stress control assessments are performed during all phases of combat operations, stability and reconstruction operations, and support operations so the term COSC may refer to any of these different types of Army operations. The COSC assessments are performed at unit and individual level. They consider the range of variables according to a model, which recognizes that biological, psychological, and social factors influence each other. The COSC assessment reviews the interaction systematically to a depth appropriate to the need. The assessment identifies which variables can be modified to improve coping or outcome. Based on these assessments COSC personnel recommend courses of action (COA) to the commander. They identify and initiate COSC interventions to improve unit effectiveness and Soldier efficiency and well-being.

1-7. Provided in Figure 1-1 is a conceptual model of stress, its mitigating and aggravating factors and potential outcomes on Soldiers and families. This model can be helpful when designing COSC interventions to improve short-term and long-term outcomes.

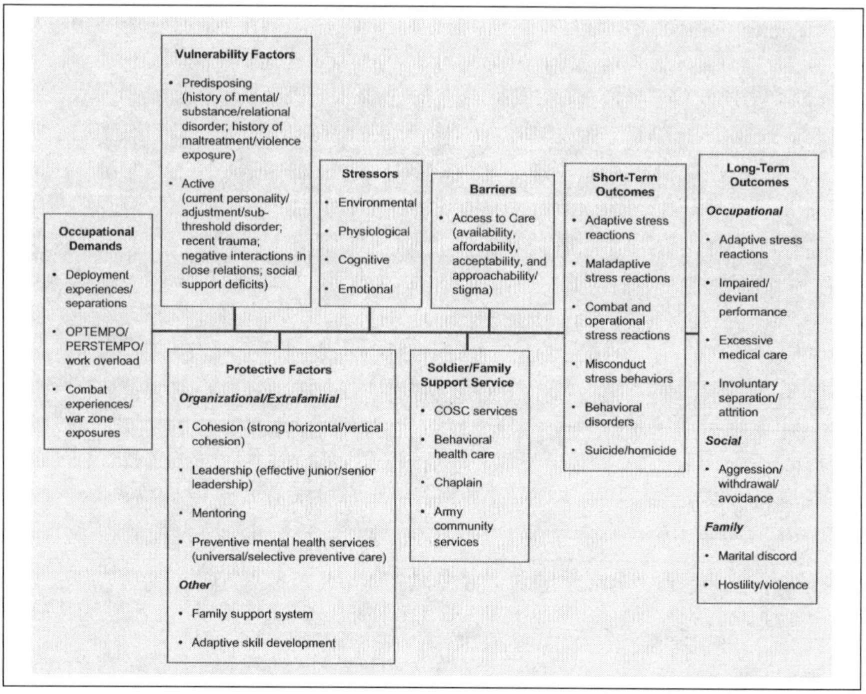

Figure 1-1. Model of stress and its potential Soldier and family outcomes

MENTAL AND PHYSICAL STRESSORS

1-8. A rough distinction between a mental and physical stressor can be made—

- A mental stressor is one in which the brain receives information about a given threat or demand, but this information results in only indirect physical impact on the body. Instead, its primary effect is to place demands on and evoke reactions from the perceptual, cognitive and/or emotional systems of the brain (such as information overload, perceived lack of control, or grief-producing losses).
- A physical stressor has a direct, potentially harmful effect on the body. These stressors may be external environmental conditions (such as temperature) or the internal physiologic demands required by or placed upon the human body (such as the need for hydration, or an immune response to a viral infection).

TYPES OF MENTAL AND PHYSICAL STRESSORS

1-9. Figure 1-2 provides examples for the two types of physical stressors (environmental and physiological) and the two types of mental stressors (cognitive and emotional). Also, as recognized in the COSC intervention model (Figure 1-1) physical stressors cause mental stressors when they result in discomfort, distraction, and threat of harm, as well as when they directly impair brain functions. Mental stressors can lead to adaptive or maladaptive stress behaviors that decrease or increase the exposure to physical stressors.

PHYSICAL STRESSORS	MENTAL STRESSORS
ENVIRONMENTAL	**COGNITIVE**
Heat, cold, wetness, dust Vibration, noise, blast Noxious odors (fumes, poisons, chemicals) Directed-energy weapons/devices Ionizing radiation Infectious agents Physical work Poor visibility (bright lights, darkness, haze) Difficult or arduous terrain High altitude	Information (too much or too little) Sensory overload or deprivation Ambiguity, uncertainty, unpredictability Time pressure or waiting Difficult decision (rules of engagement) Organizational dynamics and changes Hard choices versus no choice Recognition of impaired functioning Working beyond skill level Previous failures
PHYSIOLOGICAL	**EMOTIONAL**
Sleep deprivation Dehydration Malnutrition Poor hygiene Muscular and aerobic fatigue Overuse or underuse of muscles Impaired immune system Illness or injury Sexual frustration Substance use (smoking, caffeine, alcohol) Obesity Poor physical condition	Being new in unit, isolated, lonely Fear and anxiety-producing threats (of death, injury, failure, or loss) Grief-producing losses (bereavement) Resentment, anger, and rage-producing frustration and guilt Inactivity producing boredom Conflicting/divided motives and loyalties Spiritual confrontation or temptation causing loss of faith Interpersonal conflict (unit, buddy) Home-front worries, homesickness Loss of privacy Victimization/harassment Exposure to combat/dead bodies Having to kill

Figure 1-2. Examples of combat and operational stressors

STRESS BEHAVIORS IN COMBAT AND OTHER OPERATIONS

1-10. Combat and operational stress behavior is the generic term that is used for the full spectrum of combat and operational stress behaviors. It covers the range of reactions, from adaptive to maladaptive. Figure 1-3 provides a listing of typical adaptive and maladaptive stress reactions.

ADAPTIVE STRESS REACTIONS

1-11. Stressors, when combines with effective leadership and good peer relationships may lead to adaptive stress reactions which enhance individual and unit performance. Examples of adaptive stress reactions include—

- The strong personal trust, loyalty, and cohesiveness (called horizontal bonding), which develops among peers in a small military unit.
- Personal trust, loyalty, and cohesiveness that develops between leaders and subordinates (called vertical bonding).
- Esprit de corps is also defined as a feeling of identification and membership in the larger, enduring unit with its history and ideas. This may include the unit (such as battalion, brigade

combat team [BCT], regiment, or division) the branch (such as infantry, artillery, or military police [MP]) and beyond the branch to the US Army level.

- Unit cohesion is the binding force that keeps Soldiers together and performing the mission in spite of danger and death. Cohesion is a result of Soldiers knowing and trusting their peers and leaders and understanding their dependency on one another. It is achieved through personal bonding and a strong sense of responsibility toward the unit and its members. The ultimate adaptive stress reactions are acts of extreme courage and almost unbelievable strength. They may even involve deliberate self-sacrifice.

MALADAPTIVE STRESS REACTIONS

1-12. Combat and operational stress reaction and misconduct stress behaviors comprise the maladaptive stress reactions.

COMBAT AND OPERATIONAL STRESS REACTION

1-13. The Army uses the DOD-approved term/acronym COSR in official medical reports. This term can be applied to any stress reaction in the military unit environment. Many reactions look like symptoms of mental illness (such as panic, extreme anxiety, depression, hallucinations), but they are only transient reactions to the traumatic stress of combat and the cumulative stresses of military operations. Some individuals may have behavioral disorders that existed prior to deployment or disorders that were first present during deployment, and need BH intervention beyond the interventions for COSR.

MISCONDUCT STRESS BEHAVIORS

1-14. Examples of misconduct stress behaviors are listed in Figure 1-3. These range from minor breaches of unit orders or regulations to serious violations of the Uniform Code of Military Justice (UCMJ) and the Law of Land Warfare. Misconduct stress behaviors are most likely to occur in poorly trained undisciplined Soldiers, however good and heroic, under extreme combat stress may also engage in misconduct. Generally, misconduct stress behaviors—

- Range from minor breaches of unit orders or regulations to serious violations of the UCMJ and the Law of Land Warfare.
- May also become a major problem for highly cohesive and proud units. Such units may come to consider themselves entitled to special privileges and, as a result, some members may relieve tension unlawfully when they stand-down from their military operations. They may lapse into illegal revenge when a unit member is lost in combat. Stress control measures and sound leadership can prevent such misconduct stress behaviors, but once serious misconduct has occurred, Soldiers must be punished to prevent further erosion of discipline. Combat stress, even with heroic combat performance, cannot justify criminal misconduct.

OVERLAPPING STRESS BEHAVIORS

1-15. The distinctions among adaptive stress reactions, misconduct stress behaviors, and COSRs are not always clear. Indeed, the three categories of combat and operational stress behaviors may overlap. Soldiers with COSR may show misconduct stress behaviors and vice versa. Soldiers with adaptive stress reactions may also suffer from COSR. Finally, excellent combat Soldiers that have exhibited bravery and acts of heroism may also commit misconduct stress behaviors.

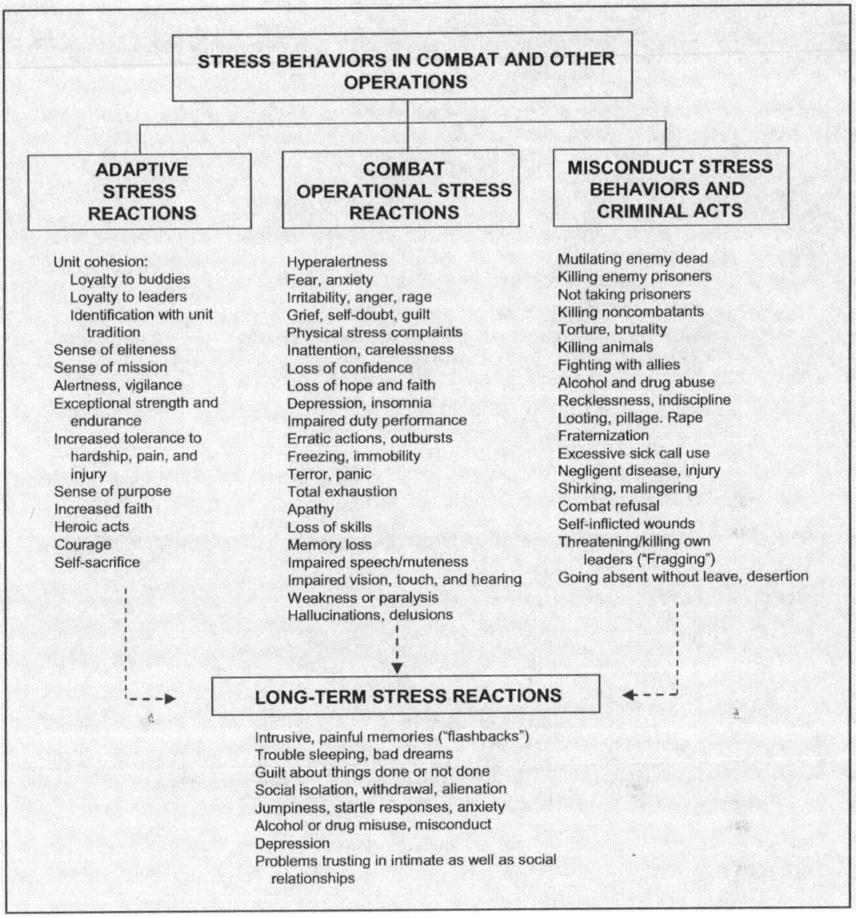

Figure 1-3. Stress behaviors in combat and other operations

1-16. It is common for stress reactions to persist or arise long after exposure to distressing events. When there is impairment in social and/or occupational functioning, a clinical assessment is warranted. Combat and operational stress control is important to sustain Army strength over the long-term and reduce the cost to society, the DOD, the Soldiers, and families.

SECTION II — PRINCIPLES AND FUNCTIONAL AREAS FOR COMBAT AND OPERATIONAL STRESS CONTROL

COMBAT AND OPERATIONAL STRESS CONTROL INTERVENTIONS

1-17. Soldier and unit readiness is best achieved through an active, prevention-focused orientation that is embodied in the principles. These principles apply to all COSC interventions or activities throughout the theater and are followed by medical personnel in all BH/COSC elements. Their application may differ based on a particular level of care and other factors pertaining to the mission, enemy, terrain and weather, troops and support available, time available, and civil considerations (METT-TC).

1-18. Preventive interventions seek to reduce the occurrence or severity of COSR and behavioral disorders, thereby sustaining Soldier and unit readiness. These interventions are tailored to the needs of the population. There are four categories of preventive interventions that include—

- Universal. Interventions targeted to the general population or an assigned AO.
- Selective. Interventions targeted to a unit or Soldier whose risk is higher than average.
- Indicated. Interventions targeted to Soldiers with COSR or indications of a potential behavioral disorder, and to units that show signs their mission effectiveness is being affected by combat and operational stressors.
- Treatment. Interventions targeted to treat and follow-up Soldiers with behavioral disorders to prevent their loss from duty.

1-19. Identify life- or function-threatening medical, surgical, or psychiatric condition as soon as possible and provide those patients emergency treatment.

COMMUNICATIONS

1-20. To maximize prevention, COSC personnel must maintain a high degree of involvement with Soldiers and leaders of supported units. Regular visits to the battalion/squadron level are essential to maintain appropriate accessibility. The COSC personnel must always be attuned to Soldiers around them and not rely on Soldiers to come to them. This means that when time permits, COSC personnel need to start conversations with as many supported Soldiers as possible.

1-21. Communication is the delivery system of COSC services and therefore is essential. Use all available means to communicate/coordinate with supported units, subordinate teams, and higher headquarters to ensure delivery of effective and timely COSC and BH services.

COMBAT AND OPERATIONAL STRESS CONTROL MANAGEMENT PRINCIPLES

1-22. The COSC management principles of brevity, immediacy, contact, expectancy, proximity, and simplicity (BICEPS). These principles apply to all COSC interventions or activities throughout the theater, and are followed by COSC personnel in all BH/COSC elements. These principles may be applied differently based on a particular level of care and other factors pertaining to METT-TC.

Note. The BICEPS principles are discussed in-depth in FM 6-22.5. This is a multiservice publication with the US Marine Corps (USMC). Due to differences in organizational structure the USMC defines the "C" in BICEPS to mean *centrality*, while the US Army defines it to mean *contact*, which will be used for the remainder of this manual.

Brevity, Immediacy, Contact, Expectancy, Proximity, and Simplicity

1-23. Using BICEPS is extremely important in the management of Soldiers with COSR and/or behavioral disorders.

Brevity

1-24. Initial rest and replenishment at COSC facilities located close to the Soldier's unit should last no more than 1 to 3 days (USMC and Navy is 3 to 4 days). Those requiring further treatment are moved to the next level of care. Since many require no further treatment, military commanders expect their Soldiers to RTD rapidly.

Immediacy

1-25. It is essential that COSC measures be initiated as soon as possible when operations permit. Intervention is provided as soon as symptoms appear.

Contact

1-26. The Soldier must be encouraged to continue to think of himself as a warfighter, rather than a *patient* or a *sick person*. The chain of command remains directly involved in the Soldier's recovery and RTD. The COSC team coordinates with the unit's leaders to learn whether the overstressed individual was a good performer prior to the COSR. Whenever possible, representatives of the unit or messages from the unit tell the Soldier that he is needed and wanted back. The COSC team coordinates with the unit leaders, through unit medical personnel or chaplains, any special advice on how to assure quick reintegration when the Soldier returns to his unit.

Expectancy

1-27. The individual is explicitly told that he is reacting normally to extreme stress and is expected to recover and return to full duty in a few hours or days. A military leader is extremely effective in this area of treatment. Of all the things said to a Soldier suffering from COSR the words of his small-unit leader have the greatest impact due to the positive bonding process that occurs. A simple statement from the small-unit leader to the Soldier that he is reacting normally to COSR and is expected back soon have positive impact. Small-unit leaders should tell Soldiers that their comrades need and expect them to return. When they do return, the unit treats them as every other Soldier and expects them to perform well.

Proximity

1-28. Soldiers requiring observation or care beyond the unit level are evacuated to facilities in close proximity to, but separate from the medical or surgical patients at the battalion aid station (BAS) or medical company nearest the Soldiers' unit. It is best to send Soldiers who cannot continue their mission and require more extensive intervention to a facility other than a hospital, unless no other alternative is possible. Combat and operational stress reactions are often more effectively managed in areas close to the Soldier's parent unit. On the noncontiguous battlefield characterized by rapid, frequent maneuver and continuous operations, COSC personnel must be innovative and flexible in designing interventions which maximize and maintain the Soldier's connection to his parent unit. See FM 6-22.5 for additional definition of proximity.

Simplicity

1-29. Indicates the need to use brief and straightforward methods to restore physical well-being and self-confidence.

1-30. The actions used for COSR control (commonly referred to as the 5 R's) involve the following actions:

- Reassure of normality.
- Rest (respite from combat or break from the work).

- **R**eplenish bodily needs (such as thermal comfort, water, food, hygiene, and sleep).
- **R**estore confidence with purposeful activities and contact with his unit.
- **R**eturn to duty and reunite Soldier with his unit.

Note. Historically the Army had used the terms proximity, immediacy, expectancy and simplicity (PIES) but began using BICEPS when it became the approved joint terminology.

NONPATIENT STATUS

1-31. To prevent Soldiers with COSR from adopting the *patient* role, these guidelines should be followed:

- Keep the Soldier in uniform and hold him responsible for maintaining Soldier standards.
- Keep the Soldier separate from seriously ill or injured patients.
- Avoid giving him medications unless essential to manage sleep.
- Do not evacuate or hospitalize the Soldier unless absolutely necessary.
- Do not diagnose the Soldier prematurely.
- Transport the Soldier via general-purpose vehicles, not ambulances.

COMBAT AND OPERATIONAL STRESS CONTROL FUNCTIONAL AREAS

1-32. Combat and operational stress control interventions and activities are organized into nine functional areas. These functional areas cover the full spectrum of BH care from preventive through clinical intervention. They are defined below and are discussed further in subsequent chapters of this manual.

UNIT NEEDS ASSESSMENT

1-33. Unit needs assessment is the systematic and frequent assessment of supported units to determine the priority and types of BH interventions required. See Chapter 4.

CONSULTATION AND EDUCATION

1-34. Consultation involves the liaison with and preventive advice to commanders, staff of supported units, and Soldiers. Education involves training in concepts and skills for increasing Soldier resilience to stress. See Chapter 5.

TRAUMATIC EVENTS MANAGEMENT

1-35. Traumatic events management blends other COSC functional areas to create a flexible set of interventions specifically focused on stress management for units and Soldiers following potentially traumatizing events (PTE). Like other functional areas, COSC providers must tailor TEM to the needs of the unit and the Soldier. See Chapter 6.

RECONSTITUTION SUPPORT

1-36. Reconstitution is extraordinary action that commanders plan and implement to restore units to a desired level of combat effectiveness commensurate with mission requirements and available resources. It transcends normal day-to-day force sustainment actions. However, it uses existing systems and units to do so. No resources exist solely to perform reconstitution. In COSC reconstitution support, COSC personnel are responsible for providing unit Soldier restoration and conducting the COSC functions. This support is provided to units following traumatic events and during reconstitution, redeployment, and transition among levels of operational tempo (OPTEMPO). See Chapter 7.

Combat and Operational Stress Control Triage

1-37. Combat and operational stress control triage is the process of sorting Soldiers with COSR and/or BH disorders based upon where they can best be managed. See Chapter 8.

Combat and Operational Stress Control Stabilization

1-38. This function provides initial management of Soldiers with severe COSR or behavioral disorders. Their safety is ensured and they are evaluated for RTD potential or prepared for further treatment or evacuation. See Chapter 9.

Soldier Restoration

1-39. Soldier restoration involves the one- to three-day management of Soldiers with COSR or behavioral disorders normally near a medical treatment facility (MTF) in close proximity to his unit. This approach uses the 5 R's discussed above. See Chapter 10.

Behavioral Health Treatment

1-40. Patients with identified behavioral disorders receive ongoing evaluation, treatment and follow-up to sustain them. This functional area implies a therapist-patient relationship, clinical documentation, and adherence to clinical standards of care. See Chapter 11.

Soldier Reconditioning

1-41. Reconditioning is an intensive program of work therapy, military activities, physical training and psychotherapy. Reconditioning programs are conducted up to seven days (or more) in the corps area. Additional reconditioning may be provided in the theater outside the combat zone (CZ). See Chapter 12.

Priority of Functional Areas

1-42. The medical company/detachment, CSC commanders, MH section leaders, senior medical commanders, and command surgeons must set priorities, coordinate their actions, and allocate resources to accomplish missions based upon the total situation/METT-TC. Needs assessments and available resources guide COSC interventions. At the unit level, the unit needs assessment helps determine which functional areas (such as consultation and education; and critical event and transition management) take priority to achieve success with the COSC mission. At the Soldier level, COSC triage helps determine which functional areas (such as consultation and education, stabilization, Soldier restoration, behavioral health treatment, and reconditioning) take priority to achieve COSC mission goals.

Chapter 2

Behavioral Health and Combat and Operational Stress Control Elements in the Theater

SECTION I — MENTAL HEALTH SECTIONS

ASSIGNMENTS

2-1. Mental health sections are located in medical companies assigned to brigades, divisions, corps, and theater-level medical units.

ARMY OF EXCELLENCE

2-2. In the Army of Excellence (AOE) divisions, the division MH section (DMHS) is organic to the main support medical company (MSMC). In the armored cavalry regiments (ACRs) and separate brigades, a MH section is assigned to the forward support medical company (FSMC) of the forward support battalion (FSB) or a medical troop. In the corps and echelons above corps (EAC), the MH sections are located in the headquarters and headquarters detachment (HHD), area support medical battalion (ASMB) and in the area support medical companies (ASMCs). In those AOE divisions that are undergoing transformation, the MH sections are assigned to each of the brigade support medical companies (BSMCs) of the brigade support battalion (BSB) and in the ASMCs from the corps ASMB.

MODULAR FORCE

2-3. In the modular BCTs, Stryker BCTs (SBCTs) and the heavy BCTs (HBCTs) MH sections are assigned to the BSMC of the BSB. At echelons above brigade (EAB) and at the division, corps, or theater levels, the MH sections are assigned to the multifunctional medical battalion (MMB).

> ***Note.*** Under the modular design, the MMB will replace the ASMB, the medical logistics battalion, and the medical evacuation battalion.

ORGANIZATIONS

2-4. The AOE divisions have only one MH section that is assigned to the MSMC of the main support battalion (MSB), which is located in the division support area (DSA). See Table 2-1 for a listing of personnel assigned to the division MH section. These personnel are deployed to the supported brigade as two-man teams comprised of one officer and one MH specialist.

Table 2-1. Division mental health staff

Division Psychiatrist (MAJ, Area of Concentration [AOC] 60W00, Medical Corps [MC])
Social Work Officer (CPT, AOC 73A67, Medical Service [MS] Corps)
Clinical Psychologist (CPT, AOC 73B67, MS)
Mental Health NCO (E7, E6, and E5, Military Occupational Specialty [MOS] 68X40/30/20)
Mental Health Specialist (Four)—(Two E4 and Two E3, MOS 68X10)

FUNCTIONS AND RESPONSIBILITIES OF ALL MENTAL HEALTH SECTIONS

2-5. All MH sections regardless of their organizational assignment are tasked with providing COSC for their supported units. In all of these units, COSC is accomplished through vigorous prevention, consultation, training, educational, and Soldier restoration programs. These programs are designed to provide BH expertise to unit leaders and Soldiers where they serve and sustain their mission focus and effectiveness under heavy and prolonged stress. The MH sections identify Soldiers with COSRs who need to be provided rest/Soldier restoration within or near their unit area for rapid RTD. These programs are designed to maximize the RTD rate of Soldiers who are either temporary impaired, have a diagnosed behavioral disorder, or have stress-related conditions. Also the prevention of posttraumatic stress disorders (PTSDs) is an important objective for brigades and EAB. The behavioral health officer (either a clinical psychologist or social work officer) and MH specialist are especially concerned with assisting and training of—

- Small-unit leaders.
- Unit ministry teams (UMTs) and staff chaplains.
- Battalion medical platoons.
- Patient-holding squad and treatment squad personnel of the medical company.

RESPONSIBILITIES

2-6. The MH section has a primary responsibility for assisting commanders with COSC by implementing the brigade combat mental fitness program. Also the MH section serves as a consultant to the commander, staff, and others involved with providing prevention and intervention services to unit Soldiers and their families. The MH section has staff responsibilities for assisting the brigade surgeon with establishing brigade policy and guidance for the prevention, diagnosis, treatment, management, and RTD of stress-related casualties. This is accomplished under the guidance and in close coordination with maneuver battalions and FSMC/BSMC physicians.

FUNCTIONS

2-7. The COSC functional areas were identified in Chapter 1. The functions of all MH sections are to support the nine functional areas except Soldier reconditioning.

UTILIZATION IN GARRISON

2-8. In garrison, BH personnel assigned to the division, brigade, or corps/EAC units continue to perform the same staff and outreach functions with supported units as they do in a field environment. An increase in the BH treatment functions may be possible as a result of consolidating BH care providers. The BH providers make available their consultation skills and clinical expertise to the Soldiers of supported units and their family readiness groups. Clinical care of family members and of Soldiers that require longer-term care beyond crisis intervention, brief treatment, and medication follow-up is the responsibility of the medical department activity (MEDDAC)/medical center (MEDCEN). The MH section personnel should focus their clinical work primarily on Soldiers with problems amenable to brief treatment.

2-9. Clinical services may be provided as part of a consolidated BH activity (as is usually established by a DMHS) or by augmenting an existing MEDDAC/MEDCEN BH staff.

2-10. The MH sections should strive to reduce referrals for BH treatment by working closely with unit leaders and chaplains to control organizational stress and rapidly identify and intervene with those Soldiers having BH disorders.

2-11. When the medical company or battalion deploys on training exercises, assigned BH personnel deploy with them to provide COSC training and support. In addition, they train to improve their own technical and tactical skills.

Note. In accordance with AR 40-216, clinical responsibilities in garrison must not interfere with participation in field and deployment exercises and maintenance of combat readiness.

MEDICAL DETACHMENT, COMBAT STRESS CONTROL

MISSION

2-12. The mission of the medical detachment, CSC (TOE 08463A000) is to provide COSC interventions and stress prevention activities to supported units in its AO. The staff augments division and brigade MH sections; provides direct support (DS) to combat brigades without organic BH officers; and provides area support in its AO. The medical detachment, CSC reconstitutes other brigade and division COSC assets. The medical detachment, CSC provides COSC interventions and activities to indigenous populations as directed in stability and reconstruction operations, humanitarian assistance, disaster relief, peace support operations, and detention facility operations. The medical detachment, CSC provides COSC interventions and activities to units in support of their readiness preparation and throughout their deployment cycle.

ASSIGNMENT

2-13. The Medical Reengineering Initiative (MRI) medical detachment, CSC (TOE 08463A000) is assigned to a corps medical brigade (MEDBDE) (TOE 08422A100), or other medical command and control (C2) element. Its teams may be further attached to a medical company, CSC (TOE 08467L000), an ASMB (TOE 08456A000), a BSMC/MSMC, or a combat support hospital (CSH). See Appendix B for the proposed future medical detachment, COSC.

EMPLOYMENT AND CAPABILITIES

2-14. The detachment headquarters is usually located in the DSA. Its teams disperse throughout the division's AO, and may extend support to corps if no CSC medical company is available. At TOE Level 1, the MRI medical detachment, CSC provides—

- A preventive section capable of dividing into four preventive teams. Each team provides mobile COSC interventions and activities.
- A fitness section capable of dividing into two fitness teams. Each fitness team provides mobile COSC interventions and activities. Each team is equipped to hold 40 Soldiers at the same time. With additional logistical support, each team can accommodate additional Soldiers under surge conditions.

STAFF RESPONSIBILITIES

2-15. The medical detachment, CSC assists the C2 headquarters (of the unit to which it is assigned or attached) regarding planning and coordination of COSC support, stress threat, mental and physical stressors, stress behaviors, principles of COSC, and implementation of COSC functional areas.

DEPENDENCY

2-16. This unit is dependent on—

- Appropriate elements of the corps for FHP, religious support, legal, finance, field feeding, personnel and administrative services support; laundry and clothing exchange; mortuary affairs support; and security of enemy prisoners of war (EPW), detainee, and US prisoner patients.
- The medical headquarters to which it is assigned/attached for FHP; medical administration, logistics, including MEDLOG; medical regulating of patients; evacuation; coordination for RTD; and unit level equipment and communications-electronics (CE) maintenance.

MOBILITY AND SECURITY

2-17. This unit is 100-percent mobile. It requires 100 percent of its organic personnel and equipment be transported in a single lift, using its organic vehicles. Upon relocation, Soldiers being held will require additional transportation. This unit is responsible for perimeter defense of its immediate operational area. However, it is dependent on appropriate elements of the corps for additional security, to include security of convoy operations. Personnel of the detachment are provided weapons for their personal defense and for the defense of their patients and/or held Soldiers.

ORGANIZATION

2-18. This 43-person MRI medical detachment (Figure 2-1) is organized into a headquarters section; a preventive section composed of four CSC preventive (CSCP) teams; and a fitness section consisting of two CSC fitness (CSCF) teams.

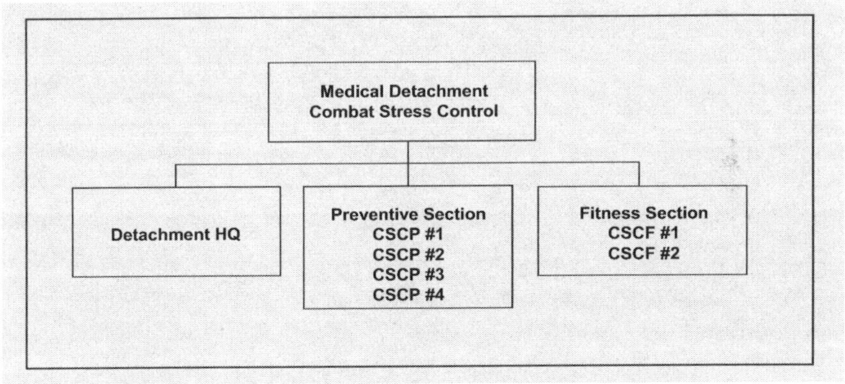

Figure 2-1. Medical Reengineering Initiative medical detachment, combat stress control (TOE 08463A000)

DETACHMENT HEADQUARTERS

2-19. The medical detachment, CSC headquarters provides C2 for the detachment. The headquarters element is responsible for planning, coordinating, and implementing COSC support for supported units. Personnel of the headquarters element provide maintenance, supply and service, and personnel administrative support. See Table 2-2 for a list of personnel assigned to the detachment headquarters.

Table 2-2. Detachment headquarters assigned personnel

Detachment Commander (LTC, AOC 60W00, MC)
Field Medical Assistant (1LT, AOC 70B67, MS)
Detachment Sergeant (E7, MOS 68X40)
Supply Sergeant (E5, MOS 92Y20)
Human Resource Specialist (E4, MOS 42A10)
Wheeled Vehicle Mechanic (E4, MOS 63B10)
Cook (E3, MOS 92G10)
Note. Detachment officers and NCOs from the preventive section and the fitness section may be assigned additional duties, which enhance the overall effectiveness of the headquarters section.

Preventive Section

2-20. The medical detachment, CSC has a 16-person preventive section. See Table 2-3 for listing of personnel assigned to the preventive section. The section can be divided into four separate CSCP teams. Three CSCP teams are normally allocated to supported maneuver brigades (one team per brigade). The fourth CSCP team may provide direct support to a SBCT, an aviation brigade, other brigade-sized units, or to corps units operating in a division's AO. Preventive section personnel may be task-organized with personnel of the fitness section into teams for specific missions. The section (and team) leader position may be held by any of the officers assigned to the section. The section's COSC interventions and activities are—

- Unit needs assessment; consultation and education; TEM; COSC triage; stabilization (emergency); and BH treatment.
- Assisting with Soldier restoration and reconditioning at the CSC detachment program. Overseeing a one- to three-day COSC Soldier restoration program in a brigade, division, or ASMC holding section or in another area suitable for Soldiers experiencing COSR and/or other stress-related disorders.

Table 2-3. Preventive section assigned personnel

Social Work Officer (Four)—(Two MAJ and Two CPT, AOC 73A67, MS)
Clinical Psychologist (Four)—(Two MAJ and Two CPT, AOC 73B67, MS)
Mental Health NCO (Four)—(E5, MOS 68X20)
Mental Health Specialist (Four)—(E4, MOS 68X10)

Fitness Section

2-21. The medical detachment, CSC has a 20-person fitness section. Personnel assigned to the fitness section are identified in Table 2-4.

Table 2-4. Fitness section assigned personnel

Psychiatrist (Two)—(LTC and MAJ, AOC 60W00, MC)
Occupational Therapist (Two)—(MAJ and CPT, AOC 65A00, Army Medical Specialty Corps [SP])
Psychiatric/BH Nurse (Two)—(MAJ and CPT, AOC 66C00, Army Nurse Corps [AN])
Occupational Therapy NCO (Four)—(Two E6 and Two E5, MOS 68W30/20N3)
Mental Health NCO (Four)—(Two E6 and Two E5, MOS 68X30/20)
Mental Health Specialist (Six)—(Four E4 and Two E3, MOS 68X10)

2-22. The fitness section is task-organized to provide COSC interventions and activities. Fitness section personnel may be task-organized with personnel of the preventive section into teams for specific missions. This section can divide into two CSCF teams. Each fitness team can deploy a four-person mobile team using the team's high-mobility multipurpose wheeled vehicle (HMMWV). One CSCF team usually collocates with a supported divisional medical company to provide mobile COSC support within a DSA and conduct Soldier restoration programs, as required. The second CSCF team augments area support to corps units in the division AO and in forward areas of the corps. The CSCF team provides staff and equipment for operating a Soldier restoration or reconditioning center. This section's COSC interventions and activities are—

- Unit needs assessment; consultation and education; TEM; COSC triage; stabilization; and BH treatment.
- Conducting Soldier restoration and reconditioning programs.
- Staffing and operating a psychiatric ward for supported CSH when this capability is required (refer to FM 8-10-14).

MEDICAL COMPANY, COMBAT STRESS CONTROL

2-23. The MRI medical company, CSC is employed in the corps and EAC. The basis of allocation is one medical company, CSC per corps or theater. The medical company, CSC is task-organized, METT-TC-dependent for stability and reconstructions operations. Medical company, CSC (TOE 08467A000) replaces medical company, CSC (TOE 08467L000). A medical company, CSC provides comprehensive preventive and treatment services to a corps and EAC during war. It provides this support to all services on an area support basis. The medical company, CSC provides DS to combat and combat support (CS) brigades, as needed. It reinforces or reconstitutes other CSC assets in the corps or divisions if required. The medical company, CSC provides COSC/BH services to indigenous populations as directed in stability and reconstruction operations, humanitarian assistance, disaster relief, and peace support operations. The comprehensive support provided by the CSC medical company entails all of the nine COSC functional mission areas. The COSC functional mission areas are identified in Chapter 1 and discussed in subsequent chapters of this publication.

CAPABILITIES

2-24. At TOE Level 1, the CSC medical company provides—

- Advice, planning, and coordination for COSC to commanders.
- Reconstitution (COSC) support for units up to division size.
- Preventive and fitness teams (four to ten personnel) for consultation, treatment services, and reconstitution support for up to battalion-sized organizations.
- Soldier restoration or reconditioning programs for up to 50 Soldiers per fitness team on an area basis.
- Deployment of COSC elements to forward areas for support of contingency operations.

ASSIGNMENT

2-25. The medical company, CSC (TOE 08467A000) is assigned to a Medical Command (US Army) (MEDCOM). Elements of this TOE may be further attached to a corps MEDBDE or to either a MMB or an ASMB.

ORGANIZATION

2-26. The medical company, CSC is organized into a headquarters section, a preventive section, and a fitness section. The company is dependent on appropriate elements of the MEDCOM or MEDBDE for administrative and MEDLOG support, medical regulating, COSR casualty delivery, and medical evacuation. The company is dependent on appropriate elements of the corps or EAC for finance, legal, personnel and administrative services, food service, supply and field services, supplemental transportation, and local security support services. When CSC medical company elements or teams are deployed to division areas, they are dependent on the division medical companies (such as MSMC or FSMC/BSMC) for patient accounting, transportation, food service, and field service support.

Employment

2-27. The medical company, CSC is employed in all intensities of conflict when a corps with two or more divisions is deployed. Task-organized COSC elements are deployed for division-sized combat operations, stability and reconstruction operations, and other contingency operations which are METT-TC-dependent. Together, the preventive and fitness sections provide all five BH disciplines. These resources are flexibly task-organized in a variety of combinations to meet the fluid COSC threat at different phases in the operations. Personnel may be quickly cross-attached from one section to another to accommodate the shifting workload and to provide reconstitution support packages. The preventive and fitness sections both organize into teams. The COSC preventive or fitness teams deployed forward of the corps boundaries in support of tactical operations come under the control of the FHP operations element in the supported units. These teams will also come under the operational control of the division or brigade CSC teams. One or more of the medical company, CSC's eight preventive teams may locate at the BSMC when deployed in DS of SBCT, HBCT, and separate brigades or ACRs.

Fitness Teams

2-28. One or more of the four (ten-person) fitness teams may reinforce ASMCs which are deployed to locations throughout the corps and EAC. These teams provide a basis for COSC prevention and intervention. The teams may conduct Soldier restoration programs at the ASMCs, as required. These teams may also be deployed forward to provide temporary augmentation/reinforcement, as required.

2-29. Based on workload, one or more of the four fitness teams, plus one or more preventive teams locate with a theater hospital where they conduct Level IV COSC reconditioning programs, as required. A hospital located in the corps rear or out of the CZ is the best location to conduct the theater COSC reconditioning program. When deployed with a hospital, these teams provide mobile consultation in the vicinity of the hospital. These teams are also prepared to restrict reconditioning programs and deploy forward in support of higher priority missions on very short notice. These teams can also augment hospital BH ward services by staffing a temporary BH ward.

2-30. The medical company, CSC is divisible into four functionally emulative increments for split-based operations and stability and reconstruction operations as assigned. Nonstandard task elements for specific missions can be organized using any combination of the preventive section and fitness section personnel to meet specific BH needs. For stability and reconstruction operations, the minimum is an officer/NCO team to supplement a brigade COSC team or a preventive module/team of two officers, one NCO, and one enlisted. These modules may be augmented with personnel from the fitness section to add additional specialty expertise.

Headquarters Section

2-31. The headquarters section provides C2 and unit-level administrative and maintenance support to its subordinate sections when they are collocated with the company. The headquarters section may also provide assistance to detached elements by making site visits if the elements are within a feasible distance for ground transportation. The CSC medical company elements normally deploy with limited maintenance capability. When these COSC elements deploy, they are dependent on the supported units for patient accounting, transportation, food service, and field services. The personnel assigned to the headquarters section are identified in Table 2-5. See Appendix C for information on the AOE L-Series TOE medical company and medical detachment, CSC.

Table 2-5. Medical company, combat stress control, headquarters section personnel

Commander (LTC, AOC 60W00, MC)
Chaplain (CPT, AOC 56A00, Chaplain [CH])
Field Medical Assistant (CPT, AOC 70B67, MS)
First Sergeant (E8, MOS 68W50)
Mental Health NCO (E7, MOS 68X40)
Chemical Operations Specialist (E5, MOS 54B20)
Patient Administrative NCO (E5, MOS 68G20)
Supply Sergeant (E5, MOS 92Y20)
Power Generator Equipment Repairer (E4, MOS 52D10)
Wheeled Vehicle Mechanic (E4, MOS 63B10)
Administrative Specialist (E4, MOS 71L10)
Patient Administrative Specialist (Two)—(E4 and E3, MOS 68G10)
Armorer (E4, MOS 92Y10)
Cook (E3, MOS 92G10)
Note. Personnel from the headquarters section are deployed with teams or task-organized COSC element, as required.

SECTION II — OPERATIONAL STRESS ASSESSMENT TEAM IN A THEATER

MISSION OF OPERATIONAL STRESS ASSESSMENT TEAMS

2-32. The operational stress assessment team (OSAT) is task-organized when needed from personnel in the research institutes of United States Army Medical Research and Materiel Command (USAMRMC). The mission of this team is to deploy with BH research expertise, equipment and supplies to provide scientifically valid needs assessment of multiple units in the theater and macro analysis of COSC stressors, stress, and Soldier needs. The source is predominantly the Walter Reed Army Institute of Research, augmented when needed with research psychologists and enlisted technicians from the US Army Aeromedical Research Laboratory, the US Army Medical Research Institute of Chemical Defense, and the US Army Research Institute of Environmental Medicine. The OSATs are USAMRMC teams that are assembled entirely from personnel in table of distribution and allowances (TDA) units and are dependent on TOE units when they deploy to a theater.

ASSIGNMENT AND CAPABILITIES

2-33. The OSAT normally is used for augmenting the senior medical command in the theater. The OSAT has the capabilities to—

- Plan, coordinate, and conduct questionnaire and interview surveys.
- Establish a database and analyze scientific-quality data from their own surveys and from standardized unit assessments conducted by unit MH section and CSC units for their local supported units.
- Provide rapid reports of findings to the command that include recommendations for policy and for COSC actions and interventions.
- Design and validate survey instruments, including ones to address new and emerging stressors and issues in the theater.
- Use USAMRMC capability to archive and analyze data for comparison with other conflicts.
- Educate and mentor COSC and other personnel in survey methodology and provide them, as well as the theater commander, the wider perspective to use in their local UNA.

- Interface with the theater surveillance capabilities of preventive medicine (PVNTMED) and occupational and environmental medicine to address the BH aspects of chemical, biological, radiological and nuclear (CBRN) exposures and threats.

DEPENDENCY

2-34. The OSAT depends on the medical headquarters or the units with which it is working for administrative support. The OSAT depends on the theater (line) command for implementation of the approved surveys.

MOBILITY AND SECURITY

2-35. The OSAT depends on the medical and theater command and the units with which they work for transportation. The OSAT personnel are issued personal weapons and protective equipment, but depend on the units they are with for protection.

COMMUNICATIONS

2-36. The OSAT depends on the medical and theater command and the units with which they work for communications. With approval of theater command, they may be allowed USAMRMC-issued phones for satellite communication for mission-essential purposes.

ORGANIZATION

2-37. The typical OSAT has three officers and two enlisted Soldiers. The officer positions are listed as research psychologists (AOC 71H), but may be filled by clinically qualified BH officers with research training and experience. The enlisted are usually MH specialists (MOS 68X) with research training and experience. Staffing may also include personnel with information management specialties. The equipment is the required personal equipment of the team members, plus data collection and automated data processing equipment appropriate to the mission.

Chapter 3

Combat and Operational Stress Control Support Operations

SECTION I — COMBAT AND OPERATIONAL STRESS CONTROL PROFESSIONAL DISCIPLINES AND PROFESSIONAL CONSULTANTS

PROFESSIONAL DISCIPLINES

3-1. There are five BH professional disciplines and two enlisted specialties that serve in support of the COSC mission. The professional disciplines and enlisted specialties were identified in Chapter 2. The professional disciplines include social work, clinical psychology, psychiatry, occupational therapy (OT), and psychiatric/BH nursing; the enlisted specialties are in BH and OT. While much of the COSC knowledge base and most of the skills are shared by all the BH personnel, each discipline brings its own perspective from its professional training, skills that can only partially be familiarized to the others, and in some cases unique credentials to conduct specific assessments and treatments. The special perspectives and skill sets of the officer and enlisted members of the COSC team are summarized in the following subparagraphs.

OFFICER AREAS OF CONCENTRATION

Social Work Officer

3-2. The social work officer (AOC 73A67) evaluates the psychosocial systems affecting the Soldier, his family, and the unit; identifies and coordinates with military and civilian support agencies (including the Army Substance Abuse Program and the Family Advocacy Program [FAP]); assesses BH disorders; provides a range of counseling and psychotherapies. Installation MTFs are responsible for BH services that are beyond the scope of COSC functional areas such as long-term clinical care, inpatient services, and alcohol treatment programs. This officer provides proactive consultation, conducts individual and group counseling, supervises Soldier restoration/reconditioning, and coordinates RTD of recovered cases. He also provides staff advice and coordinates Army and civilian social services support.

3-3. When deployed as a member of a COSC preventive team or task-organized COSC element, the social work officer's duties include—

- Evaluating psychosocial (unit and family) functioning of Soldiers with COSR and misconduct stress behavior.
- Coordinating and ensuring the return of recovered COSR and neuropsychiatric (NP) Soldiers to duty and their reintegration into their original or new unit.
- Identifying and resolving organizational and social environmental factors which interfere with combat readiness.
- Coordinating support for Soldiers and their families through Army and civilian community support agencies, when possible.
- Apprising unit leaders, primary care physicians, and others health care providers of available social service resources.
- Providing consultation to supported unit commanders and to other BH/COSC personnel regarding problem cases.
- Counseling and providing therapy for Soldiers with psychological problems.
- Conducting and supervising unit needs assessment and TEM.

3-4. The doctoral level social work officer, as outlined in DODD 6490.1, may conduct clinical evaluations for imminent risk or dangerousness.

Clinical Psychologist

3-5. A clinical psychologist (AOC 73B67) diagnoses BH disorders and a range of psychotherapies including cognitive/mental approaches, assessment for safety evaluations, and psychometric assessment.

Psychiatrist

3-6. The psychiatrist (AOC 60W00) is a physician that is credentialed to diagnose and treat BDPs. He also conducts mental assessments, makes differential diagnosis, provides a range of psychotherapies and safety evaluations, and prescribes medication.

Psychiatric/Behavorial Health Nurse

3-7. This AN officer (AOC 66C007T) provides specialized nursing services for emotionally distressed individuals and promotes COSC within the MTF and the adjacent military community. He performs liaison and consultative functions to ensure continuity of patient care. The additional skill identifier of 7T indicates that advanced skills and competencies are required for this position and that the individual may act as a consultant in clinical nursing practices.

Occupational Therapist

3-8. The occupational therapist (AOC 65A00), plans, implements, and supervises preventive and restorative OT services to enhance the occupational performance of Soldiers. This officer evaluates Soldier's performance across the spectrum of occupational areas (such as activities of daily living, work, education, leisure, and social participation). Analyze jobs and job tasks for underlying performance requisites. Identify and evaluate mental and physical stressors and stress reactions and teach prevention, adaptive coping and psychosocial skills. Apply therapeutic media in both, individual and group settings to include environmental adaptation, the therapeutic use of self, consultation, education, and occupation (purposeful activity). Match Soldiers to therapeutic and/or vocational activities based on therapeutic need and functional ability.

ENLISTED MEDICAL OCCUPATIONAL SPECIALTIES

Mental Health Specialist

3-9. The MH specialist (MOS 68X30/20) provides the perspective of the enlisted Soldier, takes the initial history and conducts mental status evaluations for supervising officer, and administers psychological tests. The MH specialist also performs counseling and may have advanced specialty training (such as drug and alcohol counseling with higher ranked individuals).

Occupational Therapy Specialist

3-10. The OT specialist (MOS 68W30/20N3) provides the perspective of the enlisted Soldier, assists the supervising OT officer in evaluating a Soldier's occupational performance; conducts initial occupational performance history interviews and mental status evaluations; observes clients to gather data as part of task performance skill assessments; and implements OT interventions under the supervision of an OT officer.

CONSULTANTS

3-11. There are two levels of COSC consultants—the theater-level consultant and the subordinate commander consultant. The theater-level COSC consultant is normally the senior COSC officer designated by the theater surgeon and approved by the theater commander. The subordinate command COSC consultants are MH officers that function as the point of contact (POC) for one or more commanders.

These consultants oversee, coordinate, and conduct COSC interventions and activities for their respective commanders. They accomplish the following tasks appropriate to their level of operation:

- Direct and report AO-wide COSC needs assessments.
- Coordinate AO-wide implementation of COSC functional areas.
- Ensure COSC standards are established and consistently applied.
- Ensure compatibility between COSC standards and hospital BH standards (theater/AO COSC consultant).
- Define COSC operational requirements.
- Recommend the AO COSC evacuation policy (theater/AO COSC consultant).
- Prepare the COSC portion of the FHP estimate and operations plan.
- Review all command medical policies affecting COSC interventions and activities.
- Evaluate the quality of COSC interventions and activities rendered in the AO.
- Make regular AO COSC consultant site visits to COSC sections and units.
- Coordinate with other COSC consultants in adjoining AOs and in the primary medical support elements to ensure unity of effort.
- Oversee AO COSC training.
- Coordinate joint and combined COSC interventions and activities as necessary.
- Plan for future COSC operations.
- Consult with the command surgeon on all COSC matters.

SECTION II — DEPLOYMENT AND EMPLOYMENT OF MENTAL HEALTH SECTIONS

MENTAL HEALTH SECTIONS

3-12. The MH section assignments, staff, functions and responsibilities were discussed in Chapter 2. The information provided in this section pertains to deployment and employment and how these sections prepare for COSC support for their units in garrison, combat operations, and stability and reconstruction operations predeployment. The involvement of MH section with support units during field and other training exercises is important. It permits BH personnel to gain a familiarity with type of units supported, how they conduct operations, their mission, and the likely stressors associated with a particular type of unit. It also allows them to interact and become familiar with unit personnel and to gain some level of trust through their interactions. Combat and operational stress control personnel should be active participants in all training exercises. Unit predeployment training must be planned when the unit is alerted for its deployment but scheduled after cross-leveling and filling of all positions is completed. Training of BH personnel will include—

- Ensuring their active participation in the normal training cycle of the BCT/ACR to which assigned.
- Training the BH officers and specialists in all COSC functional areas that support their unit.
- Working with the supported units on a frequent basis so BH personnel may gain a familiarity with the unit's personnel, capabilities, and possible stressors associated with its mission.
- Providing predeployment train-up to include refresher and operation-specific training before deployment.

PLANNING

3-13. Since planning is normally performed at battalion level, MH sections are involved in developing a COSC estimate to support the unit plan. The COSC estimate of the situation is derived from mission analysis, to include considerations for COSR, BH disorders, substance abuse, misconduct behaviors, suicide, major stressors, troop populations to be supported, and COSC augmentation requirements or assets available.

COORDINATING

3-14. The BH personnel may be involved in coordinating COSC support with supporting and supported units, as appropriate. Brigade BH personnel must be known and trusted by the leaders of the units they support. Brigade BH personnel coordinate with other BH officer and enlisted personnel pertaining to the professional disciplines associated with conducting the COSC mission. They also coordinate with other key medical and nonmedical personnel (such as health care providers, surgeons, chaplains, UMTs that assist or may be involved with the COSC mission).

DEPLOYMENT OF MENTAL HEALTH SECTIONS

Early Deployment

3-15. The MH section should be deployed early into staging areas with the lead elements of the medical company and supported units.

Command and Control

3-16. The MH section falls under the C2 of the FSMC/BSMC and may receive mission guidance from the brigade surgeon through the FSMC/BSMC commander. Tasking for COSC operational support flows through the BSMC/FSMC headquarters. When BH personnel from the medical detachment/company, CSC are attached, the MH section ensures the timely flow of operations and situation awareness information to that element.

EMPLOYMENT OF THE MENTAL HEALTH SECTIONS

3-17. Employment of the MH section is based on the type of operations being conducted. The MH section normally operates from the FSMC/BSMC located in the brigade support area (BSA). When MH personnel are deployed forward, they are dependent on supported units for rations and logistical support.

MENTAL HEALTH FUNCTIONAL AREA CONSIDERATIONS

3-18. When determining COSC support for BCT units, the BH team should review each of the BH functional areas and determine which BH functions need to be implemented for each supported unit. Also, in developing the COSC plan, the BH team should consider—

- Conducting a UNA for each supported battalion and separate company supported.
- Establishing a consultation and education/training program for supported units to include—
  - Traumatic event and transition management.
  - The need for COSC triage, stabilization, and BH treatment should be addressed as they arise.
  - Soldier restoration, which can only be accomplished within and by the BSMC holding section or with augmentation by or referral to CSC unit personnel.
  - Reconditioning services that will need to be referred to nearest location where available.

SECTION III — SUPPORT OPERATIONS CONDUCTED BY THE MEDICAL UNIT, COMBAT STRESS CONTROL

PREDEPLOYMENT ACTIVITIES FOR ALL MEDICAL UNITS, COMBAT STRESS CONTROL

3-19. The predeployment activities identified in Section II for MH sections also apply for the units in Section III that include the ASMB or MMB and medical companies and detachments, CSC. These units must also keep themselves focused on applying the nine functional areas during the predeployment phase.

Note. The MMB is the C2 organization which will replace the ASMB, the medical logistics battalion, and the evacuation battalion.

MEDICAL DETACHMENT, COMBAT STRESS CONTROL DEPLOYMENT

3-20. The medical detachment, CSC is deployed to provide COSC support whenever a division or two BCTs/regiments are deployed. It is assigned to a medical brigade or other medical headquarters for C2. Elements of the detachment may be further attached to FSMCs/BSMCs for operational control and logistical support while in support of the BCT.

MEDICAL COMPANY, COMBAT STRESS CONTROL DEPLOYMENT

3-21. The primary mission of this unit is to provide COSC support throughout a corps AO. It is assigned to either a MEDCOM/MEDBDE and may support from two or up to five divisions. They assign teams to any operational area based on troop populations without BH support. They move teams rapidly to areas of surge workload or mass casualties.

MEDICAL FORCE 2000 VERSES THE MEDICAL REENGINEERING INITIATIVE DETACHMENTS

3-22. The Medical Force 2000 (MF2K) and MRI CSC detachments conduct COSC support operations very similarly. However, in the MRI detachment, psychiatry assets are assigned to the fitness section and psychology assets are assigned forward to the prevention section as compared to the MF2K detachment, the manning strategy for psychology and psychiatry is reversed. The psychologists are assigned to the fitness section and psychiatrists are assigned to the prevention teams.

Medical Detachment, Combat Stress Control Support to a Division and Brigade Combat Teams During Combat Operations

3-23. When in support of an AOE division, a medical detachment, CSC is usually attached to the MSMC/BSMC and for a transformation division the medical detachment, CSC is attached to the MMB or MEDBDE. It can be attached to other battalion-sized units for support. It is under the operational control of the support battalion and MSMC/FSMC/BSMC and closely coordinates its augmentation support with the division surgeon and division psychiatrist. Habitual long-term relationships of medical detachments, CSC with specific divisions should be established so that COSC support is integrated as a normal function of the FHP mission. However, as an echelon above division (EAD) asset, the detachment (or its modular teams) may be task-organized and deployed to support other units or missions other than divisional units, as workloads require. The medical detachment, CSC depends on the units to which it is attached for administrative and logistics support. It may be reinforced and/or receive personnel replacements from a medical company, CSC or from another medical detachment, CSC.

Preventive Section Operations

3-24. Upon arrival to a theater, the detachment is initially attached to a higher medical headquarters for C2 and its three CSCP teams of four-person each are usually further attached. One CSCP is attached to each divisional MSMC prior to commencement of combat operations. This will allow the teams to linkup with and augment division COSC assets in support of the brigades.

- The deployment of each CSCP provides additional COSC personnel that increase COSC augmentation capabilities at the FSMC. A CSCP may routinely visit the BASs to provide consultation. The CSCP may deploy, as needed, to provide reconstitution support to units undergoing hasty or deliberate reorganization.
- In some operations, some or all of the CSCP personnel may either be attached to an ASMC or are deployed in support of a maneuver brigade or reconstitution site. It is important for all CSCP personnel to understand that they are responsible for all functional areas, and will conduct them as dictated by the mission.

Fitness Section Operations

3-25. The CSCF may locate with any Level II or above MTF in the AO. This section has tents and is equipped to operate a Soldier restoration center and can supervise Soldier restoration activities to include consultation, rest, relaxation, and treatment as necessary for speedy RTD within three days. The CSCF also provides triage and stabilization at the MSMC and consultation to other division units located in the DSA and division rear areas. All CSCF members are responsible for providing support for all COSC functional areas as dictated by the mission.

3-26. The CSCF staff may move within the AO to temporarily reinforce or reconstitute a CSCP at the brigade level, or to escort COSR casualties to the Soldier restoration center. The CSCF personnel give reconstitution support to attrited units, especially when the units return to the DSA. The CSCF supplements the division psychiatrist; brigade COSC teams, CSCPs, and chaplains and leaders in after-action debriefings. The CSCF helps integrate recovered Soldiers and new replacements into units during reorganization activities. When either the BSA or DSA is tactically too unstable to allow Soldier restoration operations, the CSCF may locate with an ASMC or a CSH and attempt to continue their support to the division from that location.

COMBAT STRESS CONTROL DETACHMENT IN CONTINGENCY OPERATIONS

3-27. For contingency operations, a task-organized CSC element from the CSC medical detachment may be deployed. A CSCP supporting a maneuver brigade may operate out of a central base of operations. The team may deploy forward to the BSA (base camps or fire bases), in response to—

- Anticipated battle.
- After-action debriefing requirements.
- Alcohol/drug problems in a unit.
- Incidents of misconduct stress behaviors.
- High incidence of COSRs.
- Unit rotation in or out of theater.

3-28. Several CSCF and CSCP from two or more CSC medical detachments may be consolidated under the command of the senior medical detachment, CSC commander to staff a central reconditioning program for the AO. This may also function as an alcohol/drug detoxification rehabilitation program. This may be referred to as a COSC center and will also provide consultation and treatment support to MP confinement facilities where misconduct stress behaviors may have led to incarceration. This could also include support for EPW and detainee confinement facilities.

STABILITY AND RECONSTRUCTION OPERATIONS

3-29. In stability and reconstruction operations, if the force deployed is smaller than a division, a medical detachment, CSC would not be required to provide COSC support. In such cases, the medical company, CSC or the medical detachment, CSC may be tasked with providing either a COSC team or a task-organized COSC element. The COSC team or the task-organized COSC element is attached to the supporting medical headquarters or to an MTF and conducts its mobile consultation mission.

MEDICAL UNITS, COMBAT STRESS CONTROL INTERFACE AND COORDINATION REQUIREMENTS

3-30. The medical units CSC must interface with its higher headquarters element and with the unit to which it is attached. The higher headquarters may be a MEDCOM, MEDBDE, or an ASMB/MMB.

- Interface between the CSC units and their higher headquarters. The medical unit CSC interfaces with its higher headquarters pertaining to its assigned mission. It provides estimates and has input to the operation plan (OPLAN). The unit receives its OPORD from the higher headquarters. Interface between the unit and higher headquarters staff elements will include the following subject areas—

- The COSC operations.
- Assignment or attachment of the medical detachment, CSC elements.
- Daily personnel and equipment status reports.
- Class VIII (medical supply) status and resupply requirements.
- Casualty feeder reports.
- Operation orders.
- Personnel replacement for the detachment.
- Medical intelligence information.
- Behavioral health/COSC consultation tasking and results.
- Maintenance requirements and requests.
- Replacement and reconstitution operations.
- Civil-military operations.
- Communications (signal operation instructions [SOI], access to message centers and nets, and transmission of COSC messages through medical, land, and other channels).
- Mass casualty plan.
- Road movement clearances.
- Tactical updates.
- Contingency operations.
- Return-to-duty and non-RTD procedures.
- Medical evacuation procedures (air and ground ambulances).
- Changes in locations of supported unit.

- Interface and coordination with unit to which attached. The headquarters of the unit to which a medical detachment, CSC is under operational control (OPCON) or attached is responsible for providing the administrative and logistical support requirements of the detachment. These requirements are normally identified in the attachment order. The higher headquarters will be identified in the attachment order and the medical detachment, CSC will be coordinated with this headquarters prior to deployment. The medical detachment, CSC must coordinate with the headquarters staff according to the tactical standing operating procedures (TSOP) of the unit of attachment. The staff shares information with the detachment commander or his representative pertaining to the threat, tactical situation, patient/COSR casualty status, and changes in FHP requirements. Coordination activities and subject area information exchange should include—
  - Command and control procedures.
  - Status of FSMCs/BSMCs and CSCPs.
  - Communications and SOI.
  - Operational support requirements.
  - Civil-military operations.
  - Soldier restoration operations.
  - Reinforcement and personnel replacement.
  - Road movement and clearances.
  - Casualty reporting and accountability.
  - Patient-holding procedures.
  - Force protection.
  - Convoy operations.
  - Geneva Conventions.
  - Detainees/EPW.
  - Improvised explosive devices (IEDs).
  - Religious and cultural considerations.

RECONDITIONING CENTERS

3-31. The medical company, CSC task-organizes CSC elements to staff separate small reconditioning centers in locations that are relatively secure. However, under some circumstances, the company may consolidate teams to establish a large reconditioning center, which supports two or three divisions. Reconditioning facilities normally locate near a CSH. See Chapter 12 for definitive information on reconditioning center operations.

Chapter 4
Unit Needs Assessment

SECTION I — PRINCIPLES AND TENETS OF UNIT NEEDS ASSESSMENT

GENERAL PRINCIPLES

4-1. The UNA is a systematic process for identifying the COSC needs of units. The UNA allows COSC personnel to identify priorities for interventions and activities and for allocating resources. The UNA is not a clinical screening to identify individuals who have or are at risk for BH disorders problems, but rather evaluates the needs of the Soldier population and leads to more effective preventive COSC activities and early interventions. The UNA allows COSC personnel to—

- Identify and describe specific areas of COSC need.
- Discover factors contributing to the needs.
- Provide an assessment of the BH training needs of Soldiers, leaders, UMT, and medical personnel within the unit.
- Develop plans to meet or improve the COSC needs of Soldiers and units through prevention and early intervention activities.

SCOPE

4-2. Unit needs assessments can be conducted at various command levels from small to large units. The COSC personnel at each level should conduct UNAs for their supported units. Larger-unit UNAs can include the composite findings and recommendations of one or more UNAs completed at subordinate levels. For example, a brigade UNA may include the results of several subordinate battalion UNAs (consolidated for confidentiality). Generally, UNAs are not conducted below the company level, though exceptional circumstances may dictate a platoon or lower UNA.

TENETS

4-3. Tenets of UNAs—

- Gain commander approval and support prior to conducting the UNA.
- Protect anonymity and ensure confidentiality of Soldiers and commanders. This includes the protection of unit identification from higher headquarters.
- Provide the commander with an unbiased assessment.
- Consider the social, political, and organizational factors of the environment.
- Ensure that information sources represent the entire unit.
- Select an assessment method that is consistent with the operational situation.
- Limit overgeneralizing the findings from one unit to another or from one time or situation to another.
- Recognize that UNAs provide population-level assessments of COSC needs, not clinical screening tools to identify individuals who may benefit from COSC interventions.
- Distinguish between what respondents report they need and what interventions are required.
- Ensure the UNA is planned and coordinated at a level commensurate with the complexity of the assessment and/or situation before starting.

SECTION II — FOCUS AND METHODS OF DETERMINING UNIT NEEDS ASSESSMENT

AREAS OF FOCUS FOR A UNIT NEEDS ASSESSMENT

4-4. A UNA involves the systematic assessment of numerous areas of Soldier and unit functioning. A typical UNA includes, but is not limited to, the following areas:

- Major stressors impacting the unit.
- Level of unit cohesion.
- Well-being of unit Soldiers.
- Soldier concern about home-front issues.
- Soldier knowledge and skill for controlling combat and operational stress.
- Soldier ideas for addressing COSC needs.
- Soldier knowledge of accessing COSC resources.
- Barriers and stigma that prevent Soldiers from accessing COSC services.
- Training needs of Soldiers, leaders, UMT, and medical personnel on topics of COSC importance (such as buddy aid, suicide awareness, or suicide prevention). See FM 6-22.5, FM 22-51, and US Army Training and Doctrine Command (TRADOC) Pamphlet 600-22 (available at: http://www.tradoc.army.mil/tpubs/pamndx.htm) for additional information on suicide awareness and prevention.

METHODS OF UNIT NEEDS ASSESSMENTS

4-5. The UNA takes advantage of all available information. Various methods can be used to assess general unit needs and to identify issues that differ among subgroups (for example, gender, rank, or race/ethnicity). The use of multiple assessment methods is recommended. These methods may include—

- Interviewing Soldiers to hear their perceptions and concerns. Information gathered during the casual conversation, although informal, may afford valuable anecdotal information (as well as developing trust and familiarity).
- Reviewing policy documents (standing operating procedures [SOPs]).
- Interviewing key unit personnel (chain of command, chaplain, and medical personnel).
- Conducting structured group interviews (focus groups or unit survey interviews).
- Administering standardized surveys and questionnaires (paper/pencil or web-based).
- Monitoring trend indicators (such as high rates of BH referrals, sick call, or misconduct, Soldier suicide, sexual assault, fratricide or disciplinary actions).
- Using multiple methods for gathering information when possible to ensure different viewpoints are considered.

Planning Considerations for Unit Needs Assessment

4-6. The UNA varies in complexity and formality depending on the purpose of the assessment and the needs of the supported unit commander. Many factors determine the complexity and formality of a UNA, which in turn influence its feasibility.

Complexity

4-7. Complexity equates to the requirements and cost to complete a UNA. Complexity is influenced by factors such as—

- The size and number of units to be assessed.
- Geographic dispersion of the units and time constraints.

Formality

4-8. Formality equates to the degree that scientific principles and methods are employed in the conduct of the UNA. An example of a UNA with low formality is one where the assessing team uses nonstructured group and individual interviews and perhaps brief questionnaires they have developed. The information is obtained from key individuals and a convenience sample of troops. That may be sufficient to quickly identify problems and make recommendations to resolve them. A UNA with high formality is one that uses professionally validated questionnaires and structured focus group interviews according to standard protocols. The data is collected from a scientifically selected sample of the larger population. The data is analyzed by standard analysis programs, so that statistically selected samples of valid comparisons can be made with other similarly sampled units in other geographical locations and across time. The OSAT expertise and assistance may be necessary for some UNA instruments and methods. In general, increasing formality increases the complexity of the UNA. The UNA requires a higher level of formality as the need for objective data and scientific precision increases (such as when results from multiple units are to be merged or compared with other UNAs).

Feasibility

4-9. Feasibility is the ability to accomplish a UNA with available resources. In developing a particular UNA, trade-offs are made to achieve an acceptable level of data quality (formality) for an affordable cost (complexity).

UNIT NEEDS ASSESSMENT PROCESS

4-10. All UNAs are conducted following a three phase plan.

PREASSESSMENT

4-11. Preassessment is an initial phase to obtain command support, determine target issues, and select appropriate methods to use.

ASSESSMENT

4-12. Assessment is the phase for gathering, integrating, and analyzing information to identify the COSC needs of the unit.

POSTASSESSMENT

4-13. The principal task of the postassessment phase is to determine the COAs to present to the commander which address the identified COSC needs. These findings are then linked to a plan of action.

WHEN TO CONDUCT UNIT NEEDS ASSESSMENTS

4-14. Unit needs assessments may be conducted at various times throughout the deployment cycle. A UNA may be conducted—

- Prior to initiation of COSC interventions and activities while COSC personnel and units establish their support relationships.
- To assess the effectiveness of COSC interventions and activities that are in the process of being conducted or that have been completed.
- At the request of supported commanders.
- To monitor trend indicators (see Paragraph 4-5).
- After serious traumatic events and significant unit transitions.
- To collect unit information for COSC planning.

4-15. Unit needs assessments should be conducted during all types of deployments, including stability and reconstruction operations, and combat operations.

Chapter 5

Consultation and Education

SECTION I — PRINCIPLES AND TENETS OF CONSULTATION AND EDUCATION

GENERAL PRINCIPLES

5-1. Combat and operational stress control consultation is defined as the transmission of information through an interactive relationship between the consultant and consultees. Education is used here in a broader sense in that it is the transmission of information by any means. Examples of consultation include providing COSC advice, coaching, training, and planning assistance. During this process, the consultant learns about the consultees and their needs, and tailors the interactions accordingly. Examples of education include distributing flyers, video and radio broadcasts, and news articles. The consultation and education functional area supports the other COSC functional areas.

CONSULTANTS

5-2. The term BH consultant is used in the general sense to describe any person performing COSC consultation or education. In addition to those theater or subordinate commander COSC consultants identified in Chapter 3, all BH/COSC personnel may serve as consultants at their level. Familiarization training with other BH disciplines enriches the ability to serve as a consultant.

CONSULTEES

5-3. Depending on the units in the area of support, broad types or categories of personnel may be consultees. These personnel are involved in recognition and control of stress as a result of their position or duty assignments. Consultation and education consultees may include—

- The command or unit surgeon and his staff.
- Staff chaplain and UMT.
- The senior commander and the senior NCO of a battalion, brigade, division, or corps.
- Staff officers and NCOs, including adjutant and personnel (S1/G1), intelligence (S2/G2), operations (S3/G3), civil-military affairs (G9), and the Judge Advocate General (JAG).
- Medical personnel such as PVNTMED teams that have missions that often complement the COSC mission.
- Company grade leaders, especially company commanders, executive officers, first sergeants (1SGs) platoon leaders, platoon sergeants, company NCOs, and Soldiers that are trained to be peer mentors.

SOLDIER-PEER MENTORS

5-4. Soldiers selected by their commanders may be trained to provide COSC help-in-place assistance for COSC information to peers. They may also serve as a POC between fellow Soldiers and the COSC and UMT teams. Selected Soldier-peer mentors with additional training could assist commanders with conducting COSC training in mission risk assessments.

TENETS FOR CONSULTATION AND EDUCATION

5-5. Consultation and education are ongoing processes that are performed across the deployment cycle and the continuum of operations. They may be provided in response to a specific request by command or recommendation of COSC personnel. Consultation and education may be provided during routine scheduled meetings such as a commander's weekly update or a leader professional development class.

5-6. Effective consultation is accomplished by the consultant's active outreach. Consultation is best conducted through recurring face-to-face contact, preferably at the consultee's location. Telephone and radio may be used to setup initial meetings and provide follow-up consultation. Consultation may be conducted one-on-one or in small groups where interaction is feasible. When necessary and feasible, audio or video teleconference may suffice. Active outreach supports the functional areas particularly COSC triage and traumatic event and transition management.

5-7. Successful consultation depends on the consultant's credibility, and the trust and familiarity established with consultees. In addition to sound professional knowledge base and clinical skills, the consultant must have military bearing and knowledge of the military (including the units, missions, vocabulary, acronyms, and skills involved). Rapport is enhanced by the demonstration of the consultant's genuine interest in the consultee and the unit. There are some situations when consultation and education are more effectively provided by the COSC officers, the NCOs, or the junior enlisted personnel.

SECTION II — CONSULTATION, EDUCATION, AND PLANNING

CONSULTATION PROCESS

5-8. The following six steps outline the process for most consultations:

- Initiate the process by introducing yourself and your capabilities to key leadership and get approval to continue.
- Assess the needs of consultees and formulate ways to address them.
- Present COA to consultee and define goals and feasibility of alternative actions.
- Implement approved COA.
- Evaluate outcomes or progress.
- Plan follow-up actions.

5-9. Consultation and education encompasses a broad range of topics, extending from prevention to treatment. The process of consultation and education assists the consultee with anticipation, identification, and control of stressors (environmental, physiological, cognitive, and emotional) and stress reactions (adaptive and maladaptive stress reactions). The following list of COSC functions, resources and their availability provides examples of consultation and education topics:

- Operation risk factors.
- Individual risk factors.
- Stress moderators.
- Unit cohesion, esprit de corps, and morale.
- Critical event management.
- Barriers to care and overcoming stigma.
- Operations planning.
- Integration of new personnel to unit.
- Trust in equipment and supporting units.
- Tough realistic training.
- Home-front stressors.
- Indicators of unit stress level.
- Combat and operational stress behaviors.
- Combat and operational stress reactions.

- Misconduct stress behaviors.
- Mental disorders.
- Suicidal and homicidal behaviors.
- Command referral processes.
- Scapegoating behaviors.
- Rumor control.
- Physical needs (such as sleep, nutrition, and hygiene).
- Social/emotional needs (such as morale, welfare, recreation push package).
- Communication with home.
- Reintegration of Soldiers into unit.
- Alcohol and substance abuse.
- Procedures for COSC triage.
- Care provider stress (such as compassion fatigue, vicarious trauma, moral dilemma).
- Long-term consequences of COSR and mental disorders.
- Medical evacuation policy and procedures.
- Leadership.
- Fear management.
- Rest and relaxation (R&R) considerations.

5-10. Additional details about some of these topics may be found in FM 22-51.

5-11. From theater staff level down to the MH sections, BH personnel are involved in the consultation and planning process for COSC support. Their support may be either through providing situational updates, developing staff estimates, or being directly involved and assisting with the development of the FHP plan. Planning starts with mission analysis. The mission analysis is the first step of the military decision making in abbreviated planning for a time-constrained situation, see FM 5-0. A part of mission analysis is based on the commander's intent and guidance. A medical battalion headquarters staff may receive addition instructions from the MEDBDE headquarters staff. During the planning process, command surgeons and medical planners (S3/G3) may seek information from the BH/COSC consultant or task medical company/detachment, CSC for COSC estimates.

MISSION ANALYSIS

5-12. The COSC planning is an essential part of every medical estimate and operations plan. Ensure that the command surgeon involves the COSC consultant in all medical planning. Mission analysis includes—

- Assessing COSC capabilities (organic and attached assets with current status and location).
- Gathering information from the commander's guidance regarding the upcoming operations. (See the S2 for intelligence, the S3 for scheme of maneuver and forces involved; the S1 for casualty estimates, personnel status issues, and units supported, and the logistics staff officer [S4/G4] for logistical considerations). Surgeons provide disease and nonbattle injuries (DNBI) estimates and status of medical units. Unit needs assessments, organic medical personnel, and UMT can provide additional information.
- Developing the COSC estimate based on the gathered information. The COSC estimate helps medical planners to anticipate demands on medical and COSC resources and to prepare accordingly. See FM 3-0, FM 5-0, and FM 8-55 for methods to develop the COSC estimate.
- Assessing limitations (specify reasons that BH assets are not available).
- Identifying specified, implied and essential COSC task in the FHP annex of the OPORD.
- Providing COSC COA.
- Presenting estimate and COA to command surgeon. Define goals and feasibility.
- Implementing approved COA. Consult with implementing COSC personnel about estimate and operational details.

- Evaluating outcomes or progress.
- Providing estimate and COA for fragmentary order (FRAGOs) and planning follow-up actions.

REHEARSAL

5-13. To achieve optimal synchronization, the FHP plan is rehearsed as an integral part of the combined arms plan at the combined arms rehearsal. Medical leaders provide input to the over all plan and develop the concept for the FHP plan. During the decision-making/orders process, they identify critical events and synchronize the FHP plans. In addition to medical locations on the CSS overlay, these plans indicate the triggers for FHP events. At the rehearsals, leaders practice their synchronized plans that include FHP.

5-14. The CSS/FHP annex of the OPORD that includes map overlays is the conclusion of the medical planning efforts; the rehearsal is the culmination of the preparation phase for an operation. The medical platoon leader has the responsibility for rehearsing FHP operations. Rehearsals are done to achieve a common understanding and a picture of how the plan will be implemented. The following leads to a successful rehearsal—

- All plans must be completed prior to the rehearsal.
- The FHP portion of the battalion rehearsals should focus on the events that are critical to mission accomplishment. A successful rehearsal ensures explicit understanding by subordinate medical personnel of their individual missions; how their missions relate to each other; and how each mission relates to the commander's plan. It is important for all medical units to understand the complete FHP concept.
- Rehearsing key FHP actions allows participants to become familiar with the operation and to visualize the "triggers" which identify the circumstances and timing for friendly actions. This visual impression helps them understand both their environment and their relationship to other units during the operation. The repetition of critical medical tasks during the rehearsal helps leaders remember the sequence of key actions within the operation and when they are executed.
- The OPORD is issued through effective troop leading procedures.

SECTION III — TRANSITION MANAGEMENT AND SUPPORT IN THE DEPLOYMENT CYCLE

SUPPORTING TRANSITION AND PHASES OF A DEPLOYMENT

5-15. Transition and phases of a deployment cycle may have different and sometimes unique stressors. These stressors may be subjects for consultation and education. Needs assessment may be necessary to redistribute COSC assets and change the means for delivering COSC activities and interventions.

5-16. The COSC personnel conduct transition (change of command) workshops, especially after relief of commanders. These workshops are normally requested by the incoming commander. The purpose of these workshops is to—

- Facilitate staff discussion of what the staff sees as the unit and staff's strong points and the areas needing more work.
- Provide the new commander the opportunity to discuss his leadership style and his expectations and set priorities for the staff.

5-17. The change of OPTEMP may include—

- Going from deployment to combat (offensive or defensive operations).
- Conducting battle handover in place by turning the battle over to another unit.
- Going from high intensity combat to stability and reconstruction operations.

PREDEPLOYMENT SUPPORT

5-18. Medical support in predeployment is mainly the responsibility of MEDDAC personnel or of a US Army Reserve (USAR) mobilization site augmentation unit (which has no BH personnel assigned).

Division and brigade COSC BH personnel are involved with their unit's Soldiers who are being treated for mental disorders, and with general COSC prevention of the unit members and families. The CSC units that are not deploying may be tasked to support predeployment COSC activities and interventions. Combat and operational stress control screening of individuals during predeployment may include—

- Secondary BH screening for Soldiers who are referred by primary care providers that flagged positive on DD Form 2795 (Pre-Deployment Health Assessment).
- Health screening of medical and BH records.
- Fitness for duty evaluations according to AR 40-501.
- Evaluations and recommendations to leadership about Soldiers who are fit for duty overall, but should not be deployed at this phase of the operation for BH reasons.

5-19. Other predeployment functions may be provided on request. These functions may include—

- Briefings, consultation and education on deployment cycle stressors and how to cope, and on specific stressors that Soldiers may encounter in the AO.
- Unit assessments at request of commanders.

MID-TOUR REST AND RECUPERATION OR EMERGENCY LEAVE

5-20. Leaders must be aware and alert for Soldiers who exhibit a need for COSC screening prior to mid-tour R&R. Other Soldiers may only require some education briefings and handout cards on handling their transition from a hostile CZ to R&R environments. In some cases, prior to emergency leave, some screening for domestic violence risk may also be necessary in any potential high-risk Soldier. This COSC screening is done for ensuring the safety of the Soldier and of others. One protocol has unit commanders referring those Soldiers that have requested leave for the purpose of addressing marital, legal, or other highly charged difficulties at home, to the BH officer. The BH officer explores the situation and the Soldier's reactions to it and checks for history with the rear detachment of the MH section, MEDDAC/MEDCEN and FAP. If the BH officer finds the Soldier to be at high risk, the commander can deny leave. Another option is to escort the Soldier to home base. At home base, he is kept by his unit's rear detachment until a meeting is held under safe conditions with the Soldier, BH, FAP personnel, his family, and the others involved in the difficulty. At this meeting the situation is defused or resolved before the Soldier is permitted to leave the rear detachment area.

REDEPLOYMENT

5-21. For Soldiers, redeployment is the process of getting orders to redeploy, embarking towards the demobilization site, or deploying to another out-of-the-theater mission. Screening of these individual Soldiers may be a requirement and have established criteria that should be used to ensure appropriate screening is conducted. The theater or specific units may require that all primary care providers who perform screening interview, postdeployment health assessment have a designated BH consultant. Available COSC/BH personnel provide secondary one-on-one BH screening for Soldiers who are referred by primary care providers because they flagged positive on DD Form 2795. Some Soldiers may require brief treatment or referral may be required as a result of the screening.

END-OF-TOUR STRESS MANAGEMENT

5-22. End-of-tour stress management is essential to reduce mental problems with Soldiers returning to their home station and/or families. This also promotes Soldiers seeking help early when problems occur. Mental health/COSC personnel consult with command on activities during redeployment and for postredeployment. Issues may include—

- Common mental disorders and potential misconduct may result when Soldiers are relieved of the stress and the focus of dangerous missions and are looking towards and preparing to go home.
- Memorial services for those Soldiers who died during the deployment.
- Recognition of outstanding performance and the equitable awarding of decorations.
- Closure ceremonies.
- Homecoming-reunion education briefing and training.

5-23. The end-of-tour debriefings (EOTDs) are like leader-led after-action debriefings, except that the time period covered extends from predeployment through to this point in redeployment. The EOTD should be conducted while the unit is still in the theater or at some place where transportation home pauses for one or more days. For each phase of the operation, the participants review what were the significant events and the problems or stressor that bothered the group (including critical events, if any). Equally important the group recalls what went well, how problems and stressors were overcome (at least by some members), and what positive memories and feelings they had and will take home.

FACILITATORS

5-24. The COSC or UMT facilitators should be invited to advise the leader and sit in if there are likely to be many negative memories and feelings that have disrupted or threaten unit cohesion. Advise leaders and troops on the importance of sustaining unit identity and contact with teammates beyond postdeployment and in the future.

POSTDEPLOYMENT

5-25. Postdeployment covers that period of time from embarking from theater, through scheduled postdeployment activities, and return to work after block leave. As with predeployment, medical support in postdeployment is mainly the responsibility of installation MTF, often augmented by a USAR mobilization site augmentation unit. Division and brigade COSC/BH personnel and CSC units that are themselves demobilizing as their units are deactivating will have limited involvement with the BH and COSC needs of fellow Soldiers. These personnel continue to provide, as needed, consultation to the command and to key personnel of the unit that they were assigned/attached. They consult with the installation MTF about BH and coordinate COSC needs that should be met. The CSC units that were not deployed may be tasked to support postdeployment COSC activities and interventions. The COSC personnel who are not demobilizing provide reintegration education. Screening of individuals postdeployment will continue and secondary screening will continue in postdeployment and be recorded on a DD Form 2796 (Post-Deployment Health Assessment). Each primary screener should have a designated BH consultant from the MEDCEN/MEDDAC. Secondary screening and treatment of referrals are provided by the installation MTF. Recent research with a returning division indicated that a secondary postdeployment screening from 90 and 120 days after returning from the theater may be necessary. This seems to be the time when persisting or delayed symptoms, problems, and perhaps impairment show need for intervention. All Soldiers should be advised about follow-up BH support resources. Soldiers that are leaving active service should also be informed about deployment-related entitlements and benefits.

READINESS PHASE OF THE DEPLOYMENT CYCLE

5-26. Division/brigade MH sections and medical units, CSC, of Active Army and Reserve Component (RC) must use their training opportunities to the fullest. During these training exercises, they should provide COSC consultation, education, and training to supported units the same as they would in an actual deployment. Mental health personnel will perform many of their critical mission functions and should not just simulate them. The management of real COSR occurs in field and garrison/home-station settings. Triage and emergency stabilization may be required. Restoration may be appropriate in some field exercises. Even reconditioning can sometimes be provided as a multiday "course" at Active Army posts where CSC units are stationed. Further, TEM and transition management are also potential major roles for organic MH sections and medical units, CSC in peacetime. The RC as well as Active Army medical units, CSC have been used extensively to respond to traumatic events and to assist units that are in the premobilization, mobilization, or postdeployment phases of their deployment cycles.

Chapter 6

Traumatic Event Management

OVERVIEW OF TRAUMATIC EVENT MANAGEMENT

TAILORING TRAUMATIC EVENT MANAGEMENT

6-1. Traumatic event management blends other COSC functional areas to create a flexible set of interventions specifically focused on stress management for units and Soldiers following potentially traumatizing event (PTE). Like other functional areas, COSC providers must tailor COSC and TEM support to the needs of the unit and Soldier.

6-2. An event is considered *potentially traumatic* when it causes individuals or groups to experience intense feelings of terror, horror, helplessness, and/or hopelessness. Guilt, anger, sadness, and dislocation of *world view* or faith are potential emotional/cognitive responses to PTEs. Studies of Soldiers in Operation Iraqi Freedom (OIF) and Operations Enduring Freedom (OEF) have shown a correlation between exposure to combat experiences and BH disorders, most particularly acute stress disorder and PTSD. The following events should be monitored as PTEs for Soldiers and units:

- Heavy or continuous combat operations.
- Death of unit members due to enemy or friendly fire.
- Accidents.
- Serious injury.
- Suicide/homicide.
- Environmental devastation/human suffering.
- Significant home-front issues.
- Operations resulting in the death of civilians or combatants.

Scope of Traumatic Events Management

6-3. In this chapter, TEM is discussed as it applies to military units and personnel to support readiness. Traumatic events management can be adapted to nonmilitary groups or individuals brought together by a PTE. Traumatic events debriefing should be conceptualized as an ongoing process and not an acute intervention. For military units and personnel, TEM is active in all phases of the deployment cycle and across the continuum of military operations.

Traumatic Events Management Functional Mission Area

6-4. Successful TEM relies on a solid foundation of other COSC functional area activities such as UNA that may require the COSC provider to—

- Establish a credible working relationship with supported unit leadership.
- Understand the unit's needs via previously conducted UNAs.

6-5. In the absence of preexisting relationships and UNA, COSC providers should secure command support and recommend a UNA as the first step in TEM. Subsequent UNAs will clarify unit responses to other TEM interventions and to ongoing unit needs.

Unit Needs Assessments

6-6. When conducting a UNA after a PTE the COSC providers need to take under consideration the following—

- Ensure a timely arrival that does not disrupt unit operations but facilitates the UNA.

- They should not interrupt or intrude on the people who are attending to the acute crisis when arriving at the unit, unless asked. Presence without intruding will gain trust information and POC.
- Understanding that the UNA may be limited by the urgency of the unit's return to action, the difficulty of data collection, and having limited resources. Sufficient knowledge must be gained to tailor the interventions to the unit before initiating them. It is better to defer the intervention to the next opportunity if the unit must return to action immediately after replenishment and/or before necessary data is collected.
- Know that the UNA is a unit-level assessment and does not substitute for individual-level screenings or COSC triage. See Chapter 4 for a detailed discussion of UNA.

Consultation and Education

6-7. The COSC provider should conduct unit leader consultation and education activities prior to a PTE. Consultation and education topics should include—

- The impact of PTE on unit and Soldier readiness.
- Common PTEs for units and Soldiers.
- An overview of TEM.
- Components of TEM.
- Normal responses to PTEs.
- Triggers to refer Soldiers for BH evaluation.
- Development of SOPs for responding to PTEs when they occur.

6-8. Preemptive consultation and education prepare unit leaders to institute TEM interventions following a PTE. The 5 R's are a good model to build on. In acute TEM interventions, leaders should consider interventions that target—

- Safety, security, and survival.
- Food, hydration, clothing, and shelter.
- Sleep.
- Medication (replace medications destroyed or lost).
- Orientation of unit/Soldiers to developing situation.
- Restoration of communication with unit, dependents, friends, and community.
- Location should be at a secure site that provides protection from ongoing threats, toxins, and harm.

6-9. Consultation and education to Soldiers should emphasize normalizing the common reactions to trauma, improving their coping skills, enhancing self-care, facilitating recognition of significant problems, and increasing knowledge of and access to COSC services. Post-PTE UNAs guide further consultation and education efforts. Combat and operational stress control providers should be aware that leaders may not have experienced the PTE in person, but have experienced the PTE through their Soldiers and may require support.

6-10. For an in-depth discussion of consultation and education, refer to Chapter 5.

Combat and Operational Stress Control Triage

6-11. The COSC provider should be prepared to provide COSC triage in the aftermath of the PTE. Consultation and education is important to ensure appropriate and timely referrals for triage. The COSC provider can offer the following guidelines in referral of Soldiers for COSC triage:

- Persistent or worsening traumatic stress reactions (such as dissociation, panic, autonomic arousal, and cognitive impairment).
- Significant functional impairments (such as role/work relationships).
- Dangerousness (suicidal or violent ideation, plan, and/or intent).

- Severe psychiatric comorbidity (such as psychotic spectrum disorder, substance use disorder, or abuse).
- Maladaptive coping strategies (such as pattern of impulsivity or social withdrawal under stress).
- New or evolving psychosocial stressors.
- Poor social support.
- Failure to respond to acute supportive interventions.
- Exacerbation of preexisting psychiatric conditions.
- Soldier request for assessment.

Stabilization

6-12. The COSC provider should be prepared to provide or coordinate stabilization services following the PTE. Pre-PTE coordination with medical unit personnel promotes safe management.

Soldier Restoration

6-13. The measures below are applicable to Soldiers with COSR following a PTE. The COSC provider should be familiar with the 5 R's and with BICEPS. In keeping with restorative efforts, the COSC provider focuses on the following measures through leadership consultation, Soldier education, and/or direct management:

- Minimizing exposure of Soldiers with COSR to further PTE.
- Reducing physiological arousal.
- Mobilizing support for those who are most distressed.
- Providing information and fostering communication and education.
- Using effective risk communication techniques.
- Proving assurance/reassurance.
- Mitigating fear and anxiety.
- Encouraging sleep hygiene.
- Reestablishing routines.
- Promoting exercise and nutrition.
- Encouraging self-paced emotional ventilation.
- Discouraging use of alcohol/substances.

Behavioral Health Treatment

6-14. Given the correlation between PTE and development of traumatic stress disorders the COSC provider must be familiar with the best current practices for evaluation and treatment. The Veterans Health Administration (VHA)/DOD Clinical Practice Guidelines website (http://www.oqp.med.va.gov/cpg/cpg.htm) offers clinicians evidence-based assessment and treatment algorithms for acute stress disorder, PTSD, and many other BH disorders.

6-15. In recent years, the use of early interventions in response to PTE has come under critical review. The focus of much of this debate is on the use of psychological debriefing (PD) and more specifically, the critical incident stress debriefings provided to individuals or groups exposed to PTEs. Current research suggests that PD/critical incident stress debriefings can be harmful to participants, while failing to reduce traumatic stress reactions or to prevent the progression to PTSD. As PD/critical incident stress debriefings undergo definitive study, COSC providers should adhere to current evidence-based best practices.

6-16. In accordance with the VHA/DOD Clinical Practice Guidelines (http://www.oqp.med.va.gov/cpg/cpg.htm) for PTSD, COSC providers should—

- Consider alternative methods to PDs for individuals affected by PTEs.
- Avoid PD as a means to reduce acute posttraumatic distress (acute stress reaction or acute stress disorder) or to slow progression to PTSD.

- Understand there is insufficient evidence to recommend for or against conducting structured group debriefings.
- Be aware that compulsory repetition of traumatic experiences in a group may be counter-productive.
- Consider group debriefings with preexisting groups (such as teams, units, emergency medical treatment [EMT] teams, coworkers, family members) may assist with group cohesion, morale, and other important variables that have not been demonstrated empirically.
- Emphasize that group participation must be voluntary.

LEADER-LED AFTER-ACTION DEBRIEFING

6-17. A leader-led after-action debriefing is lead by a platoon, squad, or team leader and is not normally conducted above platoon level. The leader-led after-action debriefing should be conducted after all missions especially when the maneuvers did not go according to plan. A leader-led after-action debriefing may even be sufficient for PTEs involving injury or death. For the leader to conduct a PD, his personnel should have received previous PDs; and normally provide peer support and validation for showing and talking about their emotional reactions during and after the debriefing. The best time to conduct this debriefing is as soon as is feasible after the team/squad/platoon has returned to a relatively safe place and members have replenished bodily needs and are no longer in a high state of arousal. Usually a well-conducted leader-led after-action debriefing is the best option to manage PTEs during a mission. The exception to this type of debriefing is when the event evoked reactions that seriously threaten unit cohesion and/or have a high likelihood of arousing disruptive behavior and emotions. In these situations the leader should ask himself the following—

- Should I conduct the debriefing?
- Should a trained facilitator be present?
- Should a request for COSC TEM be submitted for his team/squad/platoon?

CONDUCTING A LEADER-LED AFTER-ACTION DEBRIEFING

6-18. These debriefings require the leader to extend the lessons-learned orientation of the standard AAR. He uses the event reconstruction approach or has the individuals present their own roles and perceptions of the event, whichever best fits the situation and time available. See Training Circular (TC) 25-20 for definitive information on AARs. When individuals express or show emotions, the leader and the teammates recognize and normalize them; they agree to talk about them later and support the distressed Soldier through personal interactions. The group then returns to determining the facts. Lessons-learned discussion is deferred until all the facts are laid out. See FM 22-51 for additional information. The leader may provide education about controlling likely reactions or referral information at the end, depending on his knowledge and experience.

6-19. When a PTE is likely to create individual or collective guilt, distrust, or anger, the unit leader should be encouraged to request COSC assistance. Either a COSC or a UMT person trained in TEM sits in with the leader-led debriefing as a familiar and trusted friend of the unit. The COSC or UMT facilitator helps the unit/team leader rehearse and mentors the leaders on the debriefing process. During the PD, the facilitator can ask questions of the group to clarify the facts and steer the process away from divisive anger, blaming, and scapegoating toward a positive, cohesion-restoring outcome. This method is halfway between a simple leader-led after-action debriefing and a PD and is referred to as a *facilitated leader-led after-action debriefing*. The leader conducting the debriefing must be attentive to identify individuals needing COSC follow-up.

6-20. Leaders in positions above platoon level also have a role in leader-led after-action debriefing. Company commanders and 1SGs may conduct leader-led after-action debriefing with their subordinate leaders. Battalion commanders and higher may conduct leader-led after-action debriefings with their staff after distressing actions and may include subordinate leaders when time allows bringing them together.

COOL DOWN MEETINGS

6-21. A cool down meeting is referred to as an immediate, short meeting when a team or larger unit/group returns from the battlefield or other missions. These cool down meetings are held after heavy/intense battles with the enemy or a shift in the mission has occurred which is highly arousing and/or distressing. This is especially important after critical events. The cool down meeting is an informal event and occurs before the participants fully replenish their bodily needs and precedes any other activities including more COSC interventions, or return to the mission.

6-22. Personnel who coordinate and wait for the return of the unit that has been in a heavy intense battle, include leaders or supporting officers or NCOs from the command, UMT, and COSC providers. (In domestic support operations the same personnel identified above or other trained personnel from governmental or nongovernmental organizations such as the Red Cross may be waiting on the return of the unit after a PTE. These personnel may be present at the cool down meeting.)

COMPONENTS OF A COOL DOWN MEETING

6-23. Components of a cool down meeting may include—

- Assembling all of the unit personnel at a safe and relatively comfortable location for a brief period of time (about 15 minutes).
- Receiving or sharing nonstimulating beverages and convenient food (*comfort* foods if available).
- Providing personnel the opportunity to talk among themselves.
- Giving recognition and praise for the difficult mission they have completed.
- Providing information to unit personnel on where and how they will rest and replenish.
- Previewing the immediate agenda for the unit on what will happen after the cool down meeting including plans for further debriefing and/or other available stress control or morale/welfare intervention.
- Providing announcements pertaining to further preparations and expected time of return to the mission.

6-24. The COSC personnel may have a role in cool down meetings. In consultation and education, COSC consultants emphasize the value of cool down meetings and the simplicity of the components, which are easily neglected in crisis situations. When feasible, they unobtrusively attend the cool down meeting, showing "presence" while learning of the event, getting familiar with the key people, and observing anyone showing signs of distress and being available to them. If requested, they may give a very brief introduction and review of normal stress reactions that unit personnel may have in the next few hours.

Chapter 7

Combat and Operational Stress Control Support for Units Undergoing Reconstitution

UNIT RECONSTITUTION SUPPORT

7-1. Unit reconstitution support is defined in FM 100-9 as extraordinary actions that commanders plan and implement to restore units to a desired level of combat effectiveness commensurate with mission requirements and available resources. Besides normal support actions, reconstitution may include—

- Removing the unit from combat.
- Assisting the unit with external assets.
- Reestablishing the chain of command.
- Training the unit for future operations.
- Reestablishing unit cohesion.

7-2. When tasked as part of a reconstitution task force, COSC personnel are responsible for providing units with Soldier restoration, performing the COSC functions of the UNA, and providing consultation and education, as required. They provide triage, stabilization, Soldier restorations, and short time-constrained COSC treatment, when needed. Reconstitution is a time-constrained process, but Soldier restoration may be provided at the reconstitution site if several days are available. The focus of consultation and education efforts include—

- Rebuilding unit cohesion.
- Integrating new Soldiers or groups of Soldiers into the unit.
- Facilitating assumptions of command by replaced leaders.
- Facilitating the building of Soldiers' confidence in their leaders and themselves.
- Mentoring unit leaders in leader-led after-action debriefing process.
- Advising on COSC aspects of bereavement memorial services and communication with the family support group and unit families. The COSC personnel also facilitate leader-led after-action debriefings and lead or serve as observers in TEM interventions.

RECONSTITUTION PROCESS

7-3. Reconstitution of units transcends normal day-to-day force sustainment actions. Reconstitution is a total process. Its major elements are reorganization, assessment, and regeneration, in that order. All COSC personnel should be thoroughly familiar with FM 100-9.

Reorganization

7-4. Unit reorganization primarily involves a shifting of internal resources and is accomplished as either immediate or deliberate reorganization. The commander of the attrited unit decides to reorganize when required and may consider the following—

- Immediate reorganization is the quick and usually temporary restoring of degraded units to minimum levels of effectiveness. Normally, the commander implements immediate reorganization at his combat location or as close to that site as possible to meet near-term needs. The COSC personnel provide consultation with key POC in the unit (leaders, UMTs, and trauma/health care specialist, usually by telecommunication).
- Deliberate reorganization is done to restore a unit to the specified degree of combat effectiveness. Usually, more time and resources are available further to the rear. Procedures are similar to immediate reorganization except that some personnel and weapons system

replacement resources may be available, equipment repair is more intensive, and more extensive cross- leveling is possible. When used in reorganization, cross-leveling involves the movement of personnel and/or equipment between units to achieve equalization. The process is accomplished while maintaining or restoring the combat effectiveness of the units involved.

- The role of COSC personnel in deliberate reorganization may require a COSC team to deploy to the reorganization site. They assess the stressors and stress reactions and advise the commanders on supportive actions, (such as those for the reorganization of small unit-level elements), assists command with transitions and integration of new replacements.
- When the reorganization involves battalion size or larger units the process becomes more like the coordinated operation described for unit regeneration, provided below. It is likely to draw upon fitness team expertise and Soldier restoration assets from rearward as well as on the forward deployed teams.

REGENERATION

ASSESSMENT PHASES

7-5. The defining characteristic of regeneration is the massive infusion of personnel, equipment, and assistance under the directions of higher headquarters. Assessment measures a unit's capability to perform its mission. It occurs in two phases. The unit commander conducts the first phase. He continually assesses his unit before, during, and after operations. If he determines it is no longer mission-capable even after reorganization, he notifies his commander. The higher headquarters either changes the mission of the unit to match its degraded capability or removes it from combat. External elements may also have to assess the unit after it disengages. This is the second phase. These elements do a more thorough evaluation to determine regeneration needs. They also consider the resources available.

7-6. The second phase of assessment begins with an initial survey by a team sent by the higher headquarters. This team determines the status and needs of the attrited and exhausted unit as it moves to the regeneration site. Some of the key issues in estimating the COSC needs of the unit include the—

- Percentage and nature of casualties.
- Duration of operations and environmental exposure.
- Status of hydration, nutrition, and sleep.
- Loss and current effectiveness of leaders.
- Attitudes, perceptions, and level of confidence of unit survivors.
- Traumatic events.

7-7. The COSC consultants at every level must provide consultation to all command surgeons on the importance of including BH/COSC personnel in all reconstitution operations. The initial evaluation team should include a subordinate commander COSC consultant. Task-organized CSC teams normally deploy to the unit during the second phase to provide UNA, consultation, and other COSC activities and interventions. Unit needs assessment with high formality may be requested by the higher headquarters for decision-making purposes. These assessments are feasible for selected units, with prior coordination, as much of the logistical complexity is reduced by the orchestrated planning at higher command echelons.

Conducting Regeneration

7-8. Regeneration involves rebuilding a unit through large-scale replacement of personnel, equipment, and supplies. If required, it includes internally reorganizing; reestablishing or replacing essential C2 and conducting mission-essential training for the reconstituted unit. Regeneration is required when heavy losses of personnel and equipment leave a unit combat-ineffective and unable to continue its mission. The commander controlling assets to conduct regeneration decides whether to use resources for this purpose. Regeneration has two variations: incremental regeneration or whole-unit regeneration. Incremental regeneration is the massive infusion of individual personnel replacements and single items of equipment into the surviving unit elements. Whole-unit regeneration is the replacement of whole units or definable subelements, such as squads, crews, and teams. The S1/G1, S4/G4, and medical staffs coordinate the dispatch of the regeneration task-force teams. These teams occupy the reconstitution site before arrival of the exhausted unit. The COSC assets that perform the indicated initial assessment are needed as part of this task force. The reconstitution task force guides each element of the arriving units into its designated areas. The regeneration task force provides for the immediate needs of the survivors. This should include personal gear and equipment to replace lost or damaged items, food services, personnel services, maintenance teams, and medical teams to provide sick call services while organic medical personnel rest. Replacement personnel are sent to the reconstitution site. The COSC personnel assist with their assimilation into the regenerated unit.

RECONSTITUTION RESOURCE REQUIREMENTS FOR COMBAT AND OPERATIONAL STRESS CONTROL

7-9. Factors which influence the resource requirements for COSC in reconstitution (deliberate reorganization and regeneration) include the size of the unit, number of subunits which have suffered heavy casualties, the extent of emotional trauma, and time available. The focus of COSC for reconstitution support was provided in Paragraph 7-2. A guideline is provided in Table 7-1 for COSC personnel requirements. This staffing should not overly rely on organic COSC staff that will be in need of rest.

Table 7-1. Reconstitution operations guideline for combat and operational stress control personnel requirements

Size of Unit	*Personnel Required*
Company/Troop	2 to 4
Battalion/Squadron	6 to 12
BCT/Brigade Regiment	12 to 30
Division	30 to 60

Chapter 8

Combat and Operational Stress Control Triage

SECTION I — THE TRIAGE PROCESS FOR COMBAT AND OPERATIONAL STRESS CONTROL

SOLDIER TRIAGE

8-1. The COSC triage process is the sorting of Soldiers based on an assessment of their needs and capabilities, and the location where they can best be managed in keeping with BICEPS principles. Triage is applicable at every level of care. The two key components of COSC triage are assessment and disposition.

8-2. Assessment is an evaluation of the Soldier's physical and BH needs, potential medical emergencies, and other safety risks. Assessment is performed by COSC personnel according to their professional training, expertise, and standards.

8-3. Disposition is the COSC intervention plan to address the needs identified in the assessment. Disposition has two components that include—

- Determining what intervention techniques best address the Soldier's needs and functional capabilities.
- Selecting the best location where the Soldier can be managed. The personnel conducting COSC triage should consider the needs, abilities, and the safety of the Soldier. They should also consider the unit's capacity to provide COSC interventions based on its OPTEMPO mission, resources, response to prior consultations, and willingness to participate in COSC interventions.

TRIAGE ALGORITHM FOR COMBAT AND OPERATIONAL STRESS CONTROL

8-4. Like surgical triage categories, COSR also uses triage categories. Each of the COSR triage categories are discussed in detail in the below paragraphs. The COSC triage algorithm presented in Table 8-1 uses some of the triage categories.

Table 8-1. The combat and operational stress control triage algorithm

Step 1	Is this a medical emergency?				
	Yes	Refer to nearest MTF	No	Go to Step 2	
Step 2	Does the Soldier require medical/behavioral observation?				
	Yes	Go to Step 3 A	No	Does Soldier have presumptive COSR or MH disorder?	
				Yes	Go to Step 3 B
				No	Help-in-place
Step 3	A. Can the Level II MTF or CSC Soldier Restoration Center provide adequate evaluation and intervention?		B. Can the Soldier's unit support the 5 R's or other treatment interventions?		
	Yes	HOLD*	Yes	UNIT	
	No	REFER	No	Is there a suitable support unit?	
				Yes	REST
				No	HOLD*
* When deciding between two or more potential Level II MTFs or CSC Restoration Centers, refer the Soldier to the one closest to his unit that meets his COSC needs.					
Step 4	Use on subsequent triages. Has the Soldier improved after appropriate duration of intervention?				

TRIAGE CATEGORIES FOR COMBAT AND OPERATIONAL STRESS REACTION CASES

8-5. The following are triage categories that may be used for COSR cases. *Help-in-place* (HIP), *rest*, *hold*, and *refer* cases are discussed below.

HELP-IN-PLACE CASES

8-6. *Help-in-place* is used to identify those cases that do not have severe COSR or BH disorders. They are provided COSC consultation and education, as appropriate, and remain on duty. These interactions may occur in any setting (for example, dining facility, workplace, or the post exchange). Individual identifying information is not retained or documented. There is no implicit or explicit therapist-patient or therapist-client relationship in HIP interactions.

8-7. The unit identifies those cases that remain with or return to their original unit, either for full duty with their section/platoon or for light duty with extra rest and replenishment within a headquarters element. This option depends on the unit's mission, resources, and the Soldiers symptoms. Personnel performing triage must, therefore, be familiar with the unit's situation and take that into account. When the Soldier's condition improves, the Soldier and/or unit may not feel that additional triage is necessary.

Note. *Help-in-place* is a new triage category that replaces the old *duty* category and is also used for unit cases.

REST CASES

8-8. *Rest* identifies those cases that are provided rest and replenishment in a nonmedical support unit, usually one that is in support of their unit. These Soldiers do not require close medical or BH observation or treatment. They are unable to return immediately to their own unit either because their unit cannot provide an adequate environment for the 5 R's; or transportation is not available for at least a day; or the 5 R's can best be coordinated from the nonmedical support unit. This option depends on the resources and mission of the available CS/CSS units, as well as on the Soldier's symptoms. Someone must be designated to be in charge of the Soldier and ensure that the 5 R's are provided. There must be a reliable transportation link to return the Soldier to his original unit after a day or two. When the Soldier's condition improves sufficiently for him to return to his unit, the Soldier and/or the supporting unit may feel that additional triage is necessary.

HOLD CASES

8-9. *Hold* refers to those cases that require close medical/BH observation and evaluation at a Level II MTF or COSC Soldier restoration center because the Soldiers symptoms are potentially too disruptive or burdensome for any available CSS unit or element. Soldier's symptoms may be caused by a BH disorder that could suddenly turn worse and require emergency treatment. The Level II MTF or COSC Soldier restoration center must have the capability to provide the necessary medical observation, diagnostic tools, and adequate stabilization for emergency treatment. When deciding among capable Level II MTFs or COSC Soldier restoration centers, refer the Soldier to the one closest to his unit that meets his COSC needs. Assessment of closeness considers speed and reliability of transportation and back. Consider transferring to another Level II MTF or COSC Soldier restoration center with increased capabilities before changing a Soldier's triage category to *refer*. All *hold* cases will be triaged again by COSC personnel or other trained medical personnel after they have been placed in this category.

REFER CASES

8-10. *Refer* cases are similar to the *hold* cases, except that *refer* cases are too disruptive and burdensome for the MTF or the COSC Soldier restoration center that is not resourced to care for this particular case. The MTF or COSC Soldier restoration center cannot provide the necessary level of diagnostic and treatment capabilities. *Refer* cases requiring care at a COSC reconditioning center, a Level III MTF or higher levels of care will be triaged by COSC or other trained medical personnel prior to being transferred to these facilities.

TRIAGE PERSONNEL FOR COMBAT AND OPERATIONAL STRESS CONTROL

8-11. All COSC personnel participate in the triage process according to their professional training, experience, and standards. Familiarization training among BH disciplines extends the effectiveness of all COSC personnel in triage skills. Medical care providers must be mentored to use the COSC triage process. A commander providing the 5 R's and placing a Soldier in a support unit for a temporary break does not equate to COSC triage.

TRIAGE CONSIDERATIONS

8-12. Triage should be initiated when the—

- Soldier is a self-referral.
- Chaplain has referred the Soldier.
- Medical personnel have requested a COSC consultation and referred the Soldier.

- Unit member/buddy has referred the Soldier.
- Leader has requested an informal referral.
- Soldier is a command-directed referral (see DODD 6490.1 and DOD Instruction [DODI] 6490.4).
- Combat operational stress control personnel observe a Soldier's behaviors which indicate possible COSR or a BH disorder.

8-13. Factors that influence an assessment may vary in depth and duration due to several other factors. These factors may include the—

- Referral source.
- Nature of the complaint.
- Observed needs.
- Medical/psychiatric history.
- Availability of resources.
- Amount of privacy for conducting assessment.
- Operational Tempo.
- Environmental conditions.
- Professional training of the person making the assessment.
- Command interest.
- Soldier cooperation.

Note. Regardless of these factors, COSC personnel are responsible for conducting the assessment in a timely manner within professional standards of practice.

DOCUMENTATION

8-14. An assessment is documented according to AR 40-66 whenever the Soldier—

- Is diagnosed with a BH disorder.
- Has a condition (or suspected condition) requiring emergency medical evaluation or treatment.
- Is prescribed medication.
- Is assessed to be a high safety risk (for example, homicidal or suicide, cognitive impairment, substance abuse, and impulsivity).
- Requests that documentation of his assessment be made in his medical records.
- Is evacuated beyond Level II MTF for further assessment or treatment.
- Is command-referred for a BH evaluation.

TRANSFER AND EVACUATION

8-15. All COSC personnel are responsible for knowing the transfer/evacuation policies and procedures within their AO. Policies and procedural information are available through the command surgeon, medical regulating officer (MRO) of the medical command, control, communications, computers, and intelligence (C4I) headquarters or major MTF. All relevant background and/or clinical documentation must accompany the Soldier during the transportation or evacuation process.

MODE OF TRANSPORTATION

8-16. Nonambulance transport is the preferred mode of transportation for COSR and nonurgent BH cases. Examples of nonambulance transportation include the Soldier's unit vehicles, supporting supply/logistics vehicles, and nonambulance medical vehicles. Ambulances convey patient status on Soldiers and often must be reserved for medical emergency. Under the provisions of the Geneva Conventions, ambulances must be used exclusively in the performance of humanitarian duties, therefore, they cannot be used to return Soldiers to duty and transport must be provided by the supported unit.

ESCORT

8-17. Depending on a Soldier's condition, an escort (either medical or nonmedical) may be necessary to provide safety, monitoring, and accountability during transportation or evacuation. The escort should be an NCO or officer of equal or greater rank/grade as that of the escorted Soldier. Escorts must be emotionally mature, responsible, and capable of conducting their escort duties. Frequently escorts carry the Soldier's clinical documentation to the destination MTF.

FEEDBACK

8-18. Good communication is essential for effective continuity of care during the transportation/evacuation process. The Soldier's unit must be informed about his location and status throughout the process. The sending party must provide sufficient documentation about the Soldier's condition, history, and administered interventions. The receiving party must provide feedback to the sending party regarding receipt of the Soldier and his documentation.

SECTION II — PRECAUTIONS AND DIFFERENTIAL DIAGNOSTIC PROBLEMS ASSOCIATED WITH COMBAT AND OPERATIONAL STRESS CONTROL TRIAGE

PRECAUTIONS FOR COMBAT AND OPERATIONAL STRESS CONTROL TRIAGE

8-19. Medical emergencies must be identified during COSC triage. Medical emergencies consist of physical illnesses or injuries and/or BH disorders that can result in permanent injury, disability, or death. Early identification of a medical emergency avoids unnecessary delay in definitive treatment. Medical emergencies can cause emotional and/or BH changes and may resemble COSR in presentation. The following conditions and behaviors could be medical emergencies. Therefore, it is important to ensure medical examinations and disposition of Soldiers that display the following conditions—

- Psychosis.
- Mania.
- Alcohol withdrawal.
- Substance intoxication.
- Delirium.
- Suicidal gesture, attempt, or high risk for suicidal behavior.
- Catatonia.
- Significant paresis, paralysis, and/or sensory loss.

8-20. Assessing for physical illnesses or injuries is a critical part of COSC triage. The COSC personnel must always consider physical illnesses or injuries that resemble COSR or BH disorders. Physical illnesses or injuries may not reach the threshold of a medical emergency, but must be recognized and appropriately treated. Assessing for physical illnesses or injuries requires an adequate review of body systems and a quick physical examination. The examination includes vital signs, examination of head, eyes, ears, nose, throat, chest, abdomen, and extremities with simple testing of reflexes and muscle strength. In field situations (Level I and II), negative or normal findings need to be documented on DD Form 1380 (US Field Medical Card). Refer to AR 40-66 for detailed information on patient accountability and management of individual health records. Any positive findings from the physical examination must be evaluated further. If the examiner has not checked various body systems, it is not reassuring to tell a Soldier that his physical or BH complaints are only COSR. All COSC personnel should receive familiarization training on basic medical examination techniques and in documenting medical information. Whenever a physical illness or injury is suspected, personnel should consult with their medical peers for further assistance. Some cases will require direct medical examination by a physician or physician assistant (PA). The COSC personnel should not order tests or procedures that do not directly influence case management. Medical tests may promote the patient role in the mind of the Soldier. Needless tests may delay a Soldier's RTD and encourage secondary gains.

DEFER DIAGNOSIS OF BEHAVIORAL DISORDERS

8-21. During assessment, COSC personnel must always consider BH disorders that resemble COSR, but defer making the diagnosis. The COSC personnel favor this default position to preserve the Soldier's expectations of normalcy (according to BICEPS). This is also done to avoid stigma associated with BH disorders and to prevent the Soldier identifying with a patient or sick role. Deferral is also preferred because some diagnoses require extensive history collection or documentation that is unavailable during deployment situations (such as personality disorders and attention deficit hyperactive disorder). It is possible that a Soldier can have a combination of COSR, BH disorders and physical illnesses/injuries at the same time. In such cases, COSC personnel must rely on their clinical experiences, training, and consultation with peers and medical personnel to distinguish among these sometimes overlapping conditions. Physical injuries/illnesses are treated at a MTF, however, the Soldier may return for further COSC interventions and activities. Deferral of diagnosis is preferred, but diagnosis can be considered if the Soldier—

- Presents for reemerging symptoms of a previously diagnosed and/or treated BH disorder.
- Presents for refill prescription of psychotropic medication.
- Has a medical condition or BH disorder listed above in paragraph 8-19.
- Is enrolled in a reconditioning program.
- Fails to improve after having received four to five days of continuous COSC interventions and activities at hold or refer status.
- Requires individual BH treatment.
- Is referred for multiple episodes of COSR.

DIAGNOSTIC CONSIDERATIONS FOR DIFFERENTIAL DIAGNOSTIC DISORDERS

Low-Grade Environmental or Stress-Related Illnesses

8-22. Low-grade environmental or stress-related DNBI illnesses can drain the Soldier's strength and confidence. For example, chronic diarrhea and slight fever may exhaust, demoralize, and contribute to COSR among Soldiers. These conditions should be treated medically, concurrently with physical replenishment, rest, reassurance and organized activities, which restore the Soldier's confidence. If they persist in spite of rest and symptomatic treatment, a more aggressive workup and treatment may be indicated.

Dehydration

8-23. Dehydration deserves special mention because it can be very insidious. Soldiers under battlefield or heavy work conditions become extremely dehydrated without feeling thirsty. This is especially likely in CBRN equipment, or in a desert/arctic environment. Refer to FM 4-25.10 for additional information on the prevention of dehydration and FM 4-25.11 for first aid measures.

Hyperthermia

8-24. Hyperthermia (overheating) in an otherwise healthy individual often first causes mild elation and excessive energy. This may be followed by irritability, disorientation, and confusion. When core body temperature climb above 106° Fahrenheit (F) or 41° Centigrade (C), the Soldier may become belligerent, combative, and have visual hallucinations. If brain temperature rises further, the Soldier collapses and convulses in heatstroke. Refer to FM 4-25.11 for first aid measures and FM 4-25.10 for additional information on prevention and first aid measures for heat injuries.

Hypothermia

8-25. Hypothermia may cause an individual to become disoriented when core body temperature falls below 95°F (34.6°C). The person may move and speak slowly. His skin looks and feels warm, leading him to take off clothing. He becomes disoriented, then unresponsive and may appear to be dead. Hypothermia is as likely in cool or even warm, wet climates as it is in extremely cold ones. Refer to FM 4-25-10 for additional information on prevention and FM 4-25.11 for first aid measures.

Overuse Syndromes

8-26. Overuse of muscles, joints, and bones that have not been prepared for the strain of field duties can lead to persisting stiffness, pain, swelling, and orthopedic injuries. If severe, these injuries may require evacuation to a hospital for evaluation. Even if these injuries are avoided, the unfit person who overexerts has days of stiffness, aching, and weakness. Such cases are likely to develop COSR if further demands are made on them.

Rhabdomyolysis

8-27. Rhabdomyolysis is one potentially dangerous complication of severe muscle overuse (and of heatstroke or crush injuries) in which myoglobin from damaged muscle cells injures the kidneys. It can cause fatigue, seizures, muscle tenderness, and muscle aches. A warning sign is dark (tea-colored) urine, but without laboratory testing, this is not easily distinguished from the concentrated urine of dehydration. Rhabdomyolysis is a medical emergency.

Head Trauma

8-28. Concussion may stun the individual and cause amnesia, residual confusion, and/or impulsive behavior. For any case of suspected head trauma or for any case of significant memory loss (especially for a discrete period of time), check scalp, eyes, ears, nose, cranial nerve signs, and vital signs for evidence of head injury. Cases with deteriorating mental status are medical emergencies. If one pupil becomes larger than the other, it is an extreme emergency requiring immediate hospitalization. Left untreated, the condition can progress rapidly to coma and respiratory arrest within hours. If a head injury is suspected, monitor mental status and vital signs periodically, especially respiration, even though physical findings are negative. Awaken the Soldier periodically to check mental status and pupil size (allowing sufficient time to recover from any grogginess on awakening). Continuous monitoring is appropriate if there are serious concerns about the risk.

Spinal Cord Trauma

8-29. Pressure, bruising, and hematomas of the spinal cord, as well as severing of the spinal cord, can cause spinal shock, with loss of sensory and/or motor functions below the level of the injury in the affected dermatome and muscle group patterns. The loss of function may be bilateral, unilateral, or partial. These cases could be confused with paralysis or sensory loss presentations of COSR. Further manipulation of a fractured spine can worsen or make permanent the spinal cord damage. Information from the history of onset, a cautious physical and neurological examination, or complete relief of symptoms following hypnosis or strong positive suggestions could demonstrate convincingly that this is only COSR. It is best to be cautious and keep the spine immobile during care and transportation.

Postconcussion Syndromes

8-30. Postconcussion syndromes may persist weeks to months beyond the period of acute concussion. Postconcussion syndromes may include perceptual or cognitive impairment, poor impulse control, and difficulty in planning ahead. These are often accompanied by cranial nerve deficits or soft neurological signs.

Abdominal Trauma

8-31. Ruptured spleen or other intraperitoneal bleeding may cause shock. The Soldier may arrive in a fetal position and be unresponsive but have reflex guarding due to peritonitis.

Air Emboli and Focal Brain Ischemia

8-32. High blast overpressures from incoming high explosive ordnance can produce air emboli (bubbles in the blood) and focal brain ischemia (small areas in the brain which cannot get oxygen because the blood flow has been interrupted). Nuclear explosions can do this, as can high explosives when shock waves are magnified by reflection within bunkers, buildings, and trenches. Ruptured eardrums, general trauma, and evidence of pulmonary damage should be detectable. Cases may have stroke symptoms (loss of muscle strength, loss of sensation in parts of the body, and/or speech disturbances), which may be subtle and mistaken for COSR.

Laser Eye Injury

8-33. Laser range finders/target designators cause small burns on the retina if they shine directly into the eye, even at great distances and especially if viewed through optics. If the laser beam causes a small retinal blood vessel to bleed inside the eyeball, the person will see red. If blood inside the eye is confirmed on examination, the Soldier should be evacuated to a hospital with verbal reassurance that he may RTD soon. If the laser does not hit a blood vessel, the Soldier may see only flashes of light, followed quickly by some painless loss of vision. If the laser damage areas of the eye responsible for peripheral vision, the Soldier may never recognize a visual deficit. If the Soldier was looking directly at the laser source however, there will likely be a major loss of visual clarity. With simple retinal burns in the retina's periphery, most of the visual symptoms recover with hours to days of rest, reassurance, and nonspecific treatment the same as with COSR. Calm, professional treatment at each level of medical care should emphasize that the injury is not vision-threatening and the chances for some, if not total, recovery is good. Soldiers with the simple retinal burns should provide self-care to decrease the risk of assuming a patient mind-set and to promote their chances of RTD. For additional information on the management of laser eye injuries, refer to FM 8-50.

Middle Ear Injuries/Diseases

8-34. Temporary loss of hearing can be cause by a decreased acoustic sensitivity following a brief extremely intense noise (explosive) or less intense, longer duration noise. Tinnitus (ringing in the ears) can also result from acoustic nerve damage or irritation, as well as from high doses of certain drugs, such as aspirin. Distinguishing physiologic from psychogenic hearing loss may require consultation with an otolaryngologist (ears, nose, and throat specialist).

Peripheral Neuropathies

8-35. Peripheral neuropathies include compression neuropathies, which are especially likely in military settings (for example, *rucksack palsy*). Depending on severity, they may require temporary job reclassification during convalescence. As these injuries are not life-threatening, a hasty diagnosis should not precede a trial of Soldier restoration treatment.

Uncommon Endemic Neurologic Disorders

8-36. These physical diseases can first manifest with cognitive emotional and/or behavioral symptoms. A comprehensive neurological examination is required for the definitive diagnoses. Examples include—

- Guillain-Barre Syndrome manifests with muscle paralysis, usually without sensory loss, which ascends the legs and arms, then the trunk, over hours to days. It is sometimes triggered by immunizations, as might be given to troops deploying overseas. It often progresses to a life-threatening situation as the muscles of respiration become involved. This requires evacuation to EAC and continental United States (CONUS). Fortunately, recovery is usually complete, but it takes months to years.

- Multiple sclerosis is a disease that can mimic many types of COSR/BH disorders with its sometimes transitory, shifting motor, sensory, speech, and cognitive/emotional symptoms. It is made worse by stress and may be difficult to diagnose. Once confirmed, multiple sclerosis cases should be evacuated to CONUS, as should other rare, progressive diseases like Lou Gehrig's disease (amyotrophic lateral sclerosis).
- True convulsive seizure can manifest after head injury or a sublethal or chronic nerve agent exposure. These are treated with normal anticonvulsant medications. Fear of nerve agent exposure may lead some Soldiers to experience psychogenic seizures. These psychogenic seizures are also called pseudo-seizures. In addition to falling unconscious and convulsing, urinary and fecal incontinence can occur during a pseudo-seizure.

SUBSTANCE ABUSE/DEPENDENCE

ALCOHOL

8-37. Substance abuse is an example of misconduct stress behaviors and not necessarily COSR. Drug and alcohol abuse may occur in active CZs and nearby areas where use is explicitly prohibited and severely punished. Personnel performing the COSC assessment should be familiar with evaluation and treatment of substances abuse and dependency. Combat operational stress control providers should consider the following:

- Heavy habitual use of alcohol, even by otherwise capable officers and NCOs, may go unnoticed in peacetime. However, alcohol abuse it may degrade necessary mission performance demanded by combat and may result in withdrawal symptoms when access to alcohol is interrupted. Alcohol withdrawal is potential medical emergency; consultation with medical personnel is essential.
- Intoxication or withdrawal from alcohol, barbiturates, and tranquilizers may be mistaken for COSR or another BH disorder. Intoxication or withdrawal requires medical treatment. Withdrawal seizures or impending or ongoing delirium tremens need emergency medical treatment.

OVERUSE OF STIMULANTS

8-38. Stimulants may cause panic attacks, hyperactivity, mania, rage attacks, psychosis or paranoia. Cessation of amphetamines after prolonged use causes a crash characterized by extreme sleepiness, lethargy, overeating depression and suicidal thinking. This condition may require one to two weeks of hospitalization to assure safe recovery.

HALLUCINOGENIC DRUGS

8-39. Hallucinogenic drugs cause sensory distortion, panic, bizarre thoughts, and potentially dangerous behaviors. These drugs may be employed by the enemy as chemical or biological warfare agents. Phencyclidine hydrochloride (PCP) is especially problematic since it also blocks pain and tends to make those under its influence paranoid, violent, and abnormally strong. Hallucinogenic drug psychosis should not be treated with antipsychotic drugs. Physically restrain and sedate patients as necessary.

INHALATION OF FUMES

8-40. Inhalation of fumes (either by accident or as deliberate abuse) and carbon monoxide poisoning can cause disoriented, abnormal behavior. Supportive treatment and, specific antidotes/medication may be needed.

ANTICHOLINERGIC DELIRIUM

8-41. In combat, atropine delirium may occur. Soldiers are equipped with atropine injectors to use as first aid against nerve agents. Two milligrams (mg) (one atropine injector) without nerve agent challenge can cause rapid pulse, dry mouth, slightly dilated pupils, decreased sweating (hot, dry, flushed skin), and urinary retention. In some individuals, 6 mg of atropine (equal to three atropine injectors) may cause hallucination and disorientation in the absence of a nerve agent challenge. Such side effects may be more pronounced in sleep-deprived Soldiers. Overheated Soldiers are more susceptible to the atropine side effects. Atropine compounds the complications of overheating by diminishing the body's ability to lose heat through sweating. One dose (2 mg) of atropine can reduce the efficiency of heat-stressed Soldiers. Two doses (4 mg) will sharply reduce combat efficiency, and 6 mg will incapacitate troops for several hours. Some plants can also cause anticholinergic delirium when eaten.

ANTICHOLINESTERASES

8-42. A nerve agent is an anticholinesterase similar to many insecticides. Low-dose nerve agent exposure may produce miosis (pinpoint pupils) without other signs. Miosis decreases vision except in very bright light and may cause eye pain when attempting to focus. This miosis may take hours to days to improve spontaneously, depending on the degree and type of exposure. Evidence gathered from farm workers poisoned by insecticides suggests that mild personality changes, insomnia, nightmares and chronic persistent depressive symptoms may be seen even after use of an antidote. Low-dose nerve agent exposure may lower the seizure threshold of many Soldiers. True epileptic seizure cases must be distinguished from those Soldiers who may have pseudo-seizures.

BEHAVIORAL DISORDERED PATIENTS IN THE THEATER

PRIMARY BEHAVIORAL DISORDERS

8-43. Primary BH disorders (especially schizophrenic-form/schizophrenic disorder, major depression, and bipolar disorder) will continue to occur at approximately the same rate as in peacetime. Some Soldiers may hide their disorders by receiving care through civilian channels. Once in the theater they may experience a relapse or self-refer to a MTF when their medication supply is exhausted.

PERSONALITY DISORDERS

8-44. Preexisting personality disorders may make a Soldier unable to adapt to military life. However, studies have failed to show a relationship between personality disorders and the likelihood of breakdown in combat. Once Soldiers with personality disorders have developed COSR or a BH disorder, they may have greater difficulty recovering and RTD. Diagnosis should never be made in haste; diagnostic criteria must be supported with adequate historical information.

Chapter 9

Combat and Operational Stress Control Stabilization

SECTION I — EMERGENCY STABILIZATION

STABILIZATION

9-1. Emergency stabilization is the acute management of disruptive behavior resulting from COSR and/or a behavioral disorder. The disruptive behavior severely impacts unit functioning by posing a danger to self and/or others. In some cases, an underlying medical condition leads to the disruptive behavior and may present an additional threat to the Soldier's life. Emergency stabilization consists of interventions that temporarily reduce a disturbed Soldier's threat of self or others, thereby allowing further medical evaluation and/or treatment. Some behavioral disorders are associated with violent behavior, such as psychotic disorders, bipolar manic disorders, antisocial personality disorder, and borderline personality disorder. Violent behavior is also associated with disruption of brain functioning due to organic factors such as intoxication, hyperthermia, metabolic imbalance, or CBRN exposure.

9-2. The COSC triage process will be repeated throughout the emergency stabilization and will determine the disposition of the Soldier. After emergency stabilization, subsequent triage could result in an immediate RTD, transfer to a COSC Soldier restoration program for observation, or further evacuation.

METHODS USED FOR EMERGENCY STABILIZATION

9-3. Methods that may be used for emergency stabilization include—

- Providing verbal reassurance and reorientation are the best methods for controlling an agitated or disruptive Soldier. If these fail, a nonthreatening show of strength or force may suffice, or sedating medications may be offered to the Soldiers. If all other means fail to reduce the threat to self and/or others, physical restraint must be considered. Given the risk for violence, it is inadvisable to attempt subdue/restraint method one-on-one.
- Applying physical restraints is reserved for subduing and restraining agitated or disruptive Soldiers who fail to respond to safer and less restrictive forms of restraint (for example, verbal warnings or show of strength). Placing a disturbed Soldier into physical restraint increases the risk of injury to the Soldier and restraint team. Prolonged or improper application of physical restraint can cause injury to the disturbed Soldier. Given the potential for injury, it is paramount that COSC personnel receive training in proper physical restraining methods. Safe medical evacuation by ground (preferred) or air ambulance.
- Regardless of the method, the restrained Soldier must be checked frequently to guard against nerve injuries or impaired circulation, which may lead to skin ulcers or gangrene. It is important to check periodically to ensure the Soldier is not secretly escaping from restraints. The Soldier is provided verbal reassurances with positive expectations for his recovery each time he is checked.
- Chemical restraints (such as medication) can be administered to a disturbed Soldier to reduce the risk of harm to self or others. Medication may be offered to the Soldier in conjunction with verbal reassurances and reorientation. Chemical restraints may only be ordered by a medical professional who is authorized to prescribe medication when a Soldier is incompetent to make medical decisions for himself and/or when the Soldier's behavior places himself or others in danger. Once administered, medical personnel must observe for side effects and adverse reactions, and must consider administering additional medications as needed. When a Soldier is in physical restraints, medication may no longer be essential, but may serve to reduce the risk of escape, limb damage, and overheating. As a secondary benefit, once the medications reduce the

Soldier's agitation, others in the vicinity may feel safer and calmer. Before prescribing an antipsychotic medication, there are a few things to consider. First, some antipsychotic drugs may take several hours or days to take effect. Second, early administration of an antipsychotic drug may confuse the clinical picture for the next evaluator in the evacuation chain. The recommendation for most cases is to use no medication unless it is truly necessary for management.

SECTION II — COMBAT AND OPERATIONAL STRESS CONTROL FULL STABILIZATION

FULL STABILIZATION

9-4. Full stabilization is normally the mission of the Level III MTF such as a CSH specialty clinic's psychiatric service. Full stabilization goes beyond securing the safety of the Soldier and those around him. It provides a safe environment for the Soldier to receive treatment interventions, continued evaluation, and assessment for RTD potential. If RTD within the evacuation policy is not feasible, the full stabilization process helps to prepare the patient for a safe, long-distance evacuation. If a Level III MTF is not available, full stabilization may be accomplished by COSC personnel when appropriately supported or by using equipment diverted from the COSC Soldier restoration capability. Considerations for full stabilization may include—

- Conducting full stabilization for NP patients is desirable for the sake of the Soldier's future treatment and for the potential of returning some Soldiers to duty. However, full stabilization is personnel intensive with a relatively low RTD payoff. Providing only sufficient stabilization to allow evacuation from the theater may be acceptable in order to maintain the other COSC functions.
- Ensuring appropriate timely evacuation of Soldiers with NP/behavioral disorders according to the theater evacuation policy. It is preferred that full stabilization is achieved for all NP patients to facilitate appropriate and timely evacuation according to the theater evacuation policy.
- Assessing and triaging of COSR Soldiers undergoing full stabilization is an ongoing process. In subsequent triages, if a Soldier becomes stable the potential for RTD, they may be transferred to a COSC Soldier restoration or reconditioning program, or may be RTD directly.

TENETS OF FULL STABILIZATION

9-5. The COSC full stabilization includes ongoing evaluation of RTD potential. This requires assessment of mental status and performance capability overtime without excessive drug effects or limitation on activity. Contact with the Soldier's unit may be needed to get information on prior history and functioning. The further from the unit the Soldier has been evacuated, the more difficult it may be to contact the Soldier's unit. Full COSC stabilization normally takes several days. To the extent compatible with safety, the stabilization program should adhere to the principles and methods for treating COSR and behavioral disorder (such as BICEPS and 5 R's).

9-6. During full stabilization, special efforts should be made to maintain and reinforce the Soldier's identity as a Soldier. Techniques that may be helpful in maintaining the Soldier mind-set include—

- Keep Soldiers in duty uniform, not pajamas, as soon as they can be allowed.
- Maintain rank distinctions and appropriate military courtesy.
- Encourage self-care and helping behaviors.
- Engage in military work activities appropriate to the Soldiers' level of function and MOS.

9-7. Initial and ongoing assessments are essential to tailor the treatment to the Soldier's individual needs. It is essential that clinical documentation is available for these assessments. The Soldier's condition is an evolving one, and must be monitored throughout full stabilization. If assessed capable to RTD, efforts should be made to return the Soldier to duty.

9-8. Ongoing treatment and/or therapeutic modalities are essential to improving a Soldier's chances to RTD whether in theater or after evacuation. Therapeutic modalities are similar to those used on inpatient units, but must remain consistent with COSC principles. These modalities include medication, individual psychotherapy, group psychotherapy, and appropriate therapeutic occupations. Observed responses to therapeutic modalities allow informed recommendation for RTD status.

FULL STABILIZATION FACILITIES

9-9. Full stabilization is commonly conducted in the CSH. The CSH can provide more sophisticated procedures, laboratory and x-ray capabilities than are available at a medical company MTF. If a medical unit, CSC is providing emergency or full stabilization, the Soldier must be kept separate from other Soldiers in Soldier restoration or reconditioning.

9-10. If the Level III MTF (CSH) cannot provide sufficient inpatient psychiatric treatment for Soldiers requiring stabilization and preparation for evacuations, the theater/AO COSC consultant may recommend up to two temporary COA until the shortfalls resolve. He can recommend to the higher medical command that COSC personnel from one or more medical unit, CSC augment the Level III MTF psychiatric service until the caseload decrease or until replacements or additional COSC personnel are brought into theater or the MTF. Lastly, the medical C2 headquarters may direct that a COSC Soldier restoration or reconditioning assets be collocated with the Level III MTF to provide an "overflow" ward as well as augmenting staff. Definitive information is provided on the CSH NP ward staff capabilities in FM 4-02.10.

9-11. The MF2K CSH has a psychiatrist, three psychiatric nurses and a medical/surgical nurse, a social worker, nine mental health specialists and an OT sergeant at full strength. It has a modular hospital tent to provide a 20-bed hospital ward. The MF2K hospitals were designed to be at one location, but may follow the MRI doctrine of splitting off a portion to a remote site, taking with it some of the psychiatry service personnel. One portion would not have the psychiatry ward module, and would need to function similarly to the MRI CSH psychiatry service.

9-12. The MRI CSH has a psychiatrist, a psychiatric nurse, and two MH specialists. The service does not have an organic psychiatric ward where NP patients are admitted. There is normally an intermediate care ward with medical ward staff designated to receive NP patients. Some of these patients seen by the NP clinic will already be stable, and are being evaluated and prepared for RTD or evacuation. Most patients who are admitted for emergency or full stabilization will be on the designated ward, although a few with serious medical complications could require the intensive care ward. If the hospital has a high census of medical and surgical patients, most of the NP patients will need to occupy beds on the intermediate care ward. The MRI hospital may be augmented by staff from a medical company/detachment, CSC to operate a NP ward, as required.

9-13. Full stabilization facilities in theater can be categorized into two types:

- Mobile facilities use general purpose (GP) large or medium tents as used in medical units, CSC and can also be available to a CSH or a tent, expandable, modular, personnel (TEMPER) tent as used in a CSH.
- Fixed facilities use buildings that were previously hospitals or buildings converted to hospitals.

9-14. Facilities used for full stabilization include tents, Deployable Medical Systems (DEPMEDS) TEMPER tents, and fixed facilities. The adaptation of these facilities have both advantages and disadvantages that include—

- The principal advantage of the (hospital) TEMPER tents, as assembled into DEPMEDS hospital, is their climate control capability. This may be a significant safety advantage for treating seriously disturbed patients in restraints with high-dose medication, which can disrupt body's ability to regulate normal body temperature. The TEMPER and standard tents both pose greater problems for security than do fixed facilities. The staff may, therefore, have to rely more than is ideal on physical restraints and medications for sedation of some cases. Blankets or screens can be used to isolate or segregate problem patients from others. Such partitions reduce mental contagion but provide little true protection.

- Standard hospital beds are on high, lightweight metal legs with wheels. For full stabilization purposes, these should be replaced with standard low, stable cots to hold strong, agitated patients in restraints. The cots also make a more "military" setting and can be used as seats for group activities.
- When feasible, it is best to have a separate "closed" (high security) and "open" (moderate/minimal) security area. The latter could be a standard GP large tent (the same as those of the minimal care wards [MCWs] located close to the TEMPER tent of the official NP ward). The specialists (MOS 68WM6 [practical nurse] and 68W [health care specialist]) of the MCW could be given on-the-job training in supervision and military group activities for the moderate/minimum security cases if the NP staff is too small.

Chapter 10

Combat and Operational Stress Control Soldier Restoration

SECTION I — SOLDIER RESTORATION

SOLDIER RESTORATION PROGRAM

10-1. Soldier restoration is normally a 24- to 72-hour (1- to 3-day) program in which Soldiers with COSR receive treatment. Soldier restoration is accomplished using the principles of BICEPS and the 5 R's as discussed in the previous chapter. The 5 R's are tailored to the needs of the Soldier. Soldier restoration is conducted as close to the Soldier's unit as possible. Soldier restoration can be conducted by medical units/elements throughout the theater with the assistance of organic and/or augmented COSC personnel. The medical company/detachment, CSC is staffed and equipped to establish Soldier restoration programs. Soldier restoration may also be accomplished at a Level III MTF (CSH). The tenets remain the same but implementation may differ in the duration of Soldier restoration and the specialized skill and knowledge of available providers. There are three subcategories of Soldier restoration (referred to as lines of Soldier restoration) based on location and available resources that are discussed in Section II of this chapter.

SOLDIER RESTORATION PROCESS

10-2. The process of Soldier restoration involves several steps that include screening, assessment, and interventions, reintegration/coordination, and movement of Soldiers. A discussion of each of the steps in the process is provided below.

SCREENING

10-3. Adequate medical screening and treatment as necessary must be done before the Soldier begins the Soldier restoration program. Minor medical conditions can be treated during routine sick call. Soldiers entering a Soldier restoration program should be only those *hold* cases that require continuous medical and/or BH evaluation and observation for 24 hours or more. The criteria for *hold* cases are discussed under COSC triage in Chapter 8. Tending to and restoring physiological status (such as sleep and hydration) is a priority. Some Soldiers who need Soldier restoration will also have a mental disorder. Treatment for the mental disorder may continue or be initiated during the Soldier restoration process.

ASSESSMENT AND INTERVENTION

10-4. Initial assessment and subsequent COSC interventions depends upon the severity of the COSR and to what extent the reactions interfere with the Soldier's ability to function. More thorough interviews are conducted only after the Soldier's physiological status has been restored.

REINTERGRATION

10-5. Coordination is required to assist Soldiers with reintegration back into their units. Soldiers are reassessed regularly. When a Soldier experiencing COSR begins to improve, the COSC interventions shift toward the reintegration of the Soldier back into his unit. The COSC personnel must work with the COSC consultant of their area or other resources to assist with this reintegration.

MOVEMENT

10-6. Movement of Soldiers from one line of Soldier restoration to another that has greater capabilities or security may be required. When movement of these Soldiers is required, vehicles used to transport them should be nonambulance vehicles, if possible, and accompanied by escorts from their unit. If nonmedical unit members are not available to perform escort duties, medical augmentation may be required.

PRINCIPLES AND PROCEDURES OF SOLDIER RESTORATION

10-7. Initial Soldier restoration begins as close to the Soldier's unit as possible, normally near an MTF where the Soldier can get a reprieve from extreme stress but at the same time can be close to his unit. Normally, Soldier restoration facilities are established in the BSA near or adjacent to the FSMC/BSMC Level II MTF. Soldier restoration is not feasible at locations that are consistently under artillery, air, or direct-fire attack, but ideally are still within the sound of the artillery or other reminders of battle. If there is potential for attack, there must be cover and defensive positions. The location should not be one from which a move is likely within 24 hours. If there is a significant possibility of a move, only those cases that can participate actively in the move with minimal supervision should be managed at this location. The specific site of the Soldier restoration facility should be adjacent to the medical company's area and be out of the immediate (close) sight of the medical triage and treatment areas

REASSURANCE

10-8. Immediate reassurance is given to the Soldier with COSR beginning with the COSC triage as discussed in Chapter 8. Tell the Soldiers that they are temporarily joining the unit, not as patients, but as Soldiers who need a few days to recover from COSR. Emphasize that COSR is a normal response to extremely abnormal conditions. Rapid recovery is also normal and RTD is expected. Reassure the Soldiers about safety and what to do in the event of an attack or march order. Lastly, orient the Soldiers to the Soldier restoration program.

STRUCTURED MILITARY ENVIRONMENT

10-9. An emphasis should be placed on maintaining a highly structured military unit environment and schedule of activities in order to keep the Soldier from adopting a patient role. Assign the Soldier to a squad under supervision of a specific squad leader. The squad leader may be a CSC unit/MH section member or a member of the BSB subordinate unit. The squad leader may also be one of the Soldiers there for Soldier restoration, if their condition allows. The newly arrived Soldier is assigned to a tent that has been designated/erected for the Soldier restoration program. Soldiers with more dramatic COSR or BH symptoms should be temporarily quartered separately from other Soldiers receiving Soldier restoration. In remaining consistent with the principle of treating Soldiers with COSR as Soldiers instead of patients, it is recommended that Soldier restoration facilities do not display the Red Cross as displayed on the MTF. This could affect their protection under the Geneva Conventions. See FM 4-02 for information on the Geneva Conventions.

SUPPORT THE SOLDIER'S MILITARY IDENTITY

10-10. Maintain appropriate rank distinctions, titles, and military courtesies within the confines of the tactical situation. Expect the Soldier to maintain military bearing and personal appearance. They should be in duty uniform. Conduct basic Soldiering skills. Do not take personal possessions away from the Soldier. This includes weapons, unless there is significant concern for the Soldier's safety. If the Soldier arrives with a weapon, SOPs of the AO will guide whether or not the weapon is secure for the Soldier. Also, encourage the unit to maintain contact with its Soldier.

REPLENISHMENT OF PHYSIOLOGICAL STATUS

10-11. Get the Soldier under shelter and cool down if overheated, warm up if cold, and dry off if wet. Providing hot beverages and/or soup will also assist with restoring body temperature. Replenish hydration with palatable beverages and meals. Unless the Soldier is totally exhausted, institute some personal hygiene. If the Soldier needs uniforms or equipment, coordinate with the BSB S4 or supporting logistic element. Restorative sleep should be as normal as possible. Soldiers are typically able to fall asleep when reassured of safety, and do not routinely require medication to induce sleep. They should be informed that they may awaken with vivid and frightening dreams, and be instructed on quick relaxation techniques to go back to sleep. If those measures are insufficient, one-time medication with a sedative/hypnotic may be considered, coordinated, and prescribed by a physician or PA from the supporting medical company. The duration of sleep should be sufficient to make substantial progress in repaying the sleep debt. It should also begin the process of restoring a reasonable sleep/wake cycle that is consistent with the Soldier's duties in his unit.

RESTORE CONFIDENCE

10-12. Restoring a Soldier's confidence may include—

- Providing regularly scheduled formations to keep Soldiers informed of daily activities and the tactical situation, including information about their unit, when available.
- Providing therapeutic occupations that are based on the Soldier's current functional ability. The OT (or COSC team members under the guidance of the OT) selects therapeutic occupations that support the Soldier's military identity and enhance the Soldier's sense of competence. Therapeutic occupations may include activities of daily living (such as physical fitness or uniform maintenance); educational activities (selected common and collective Soldier task training or life skills training); work/productive activities (militarily relevant tasks such as vehicle maintenance or site maintenance); leisure/recreational activities (enjoyable, relaxing activities, games, and sports); social participatory activities (cooperative/competitive sports, games, ceremonies, or celebrations). Therapeutic occupations provide the casualty with a challenge but afford successful performance that shows the Soldier that he is still capable and competent. This realization plays the dominant role in restoring the Soldier's sense of confidence, functional capacity, and ability to RTD.
- Training and teaching Soldier on methods for managing excessive stress.
- Facilitating factional review of precipitating event that has caused the Soldier's COSR through talking (ventilating) and coaching. Combat and operational stress control personnel (or health care providers trained by COSC personnel) facilitate these discussions. They help Soldier restore perspective with questions and coaching. The factual review (debriefing) is often best done individually unless several Soldiers experienced the same event.

SECTION II — LINES OF SOLDIER RESTORATION

FIRST-LINE SOLDIER RESTORATION

10-13. First-line Soldier restoration is usually provided at the BSMC/FSMC in the BSA. It is provided by personnel organic to the BSMC/FSMC usually assisted and supervised by COSC personnel organic to the medical company and/or from CSC unit personnel. This also applies to a cavalry regiment. Because of the high mobility of the BSMC/FSMC, Soldier restoration will often be a 24- to 72-hour process. First-line Soldier restoration also may occur in relative secure locations in division, corps, and theater AOs. First-line Soldier restoration in the division is provided near and supported by a Level II MTF such as a MSMC with its organic MH and/or minimal augmentation from a CSC unit. First-line Soldier restoration in corps and theater is provided by ASMCs with responsibility for their AO. It is provided by COSC personnel organic to the medical company and/or from CSC unit personnel. The MSMC and ASMCs will be relatively unlikely to move on short notice so Soldier restoration up to 72 hours should be feasible. Support provided for Soldier restoration is a responsibility of the MH sections of supporting medical companies. Tasking for support of Soldier restoration programs is accomplished through the command surgeon.

SECOND-LINE SOLDIER RESTORATION (SOLDIER RESTORATION CENTER)

10-14. Each of the first-line Soldier restoration locations listed above should be backed up by a second-line Soldier restoration capability at a location that is less likely to have to move on short notice and has more COSC capabilities. Soldier restoration centers may be located near the MSMC, ASMC, or with a CSC medical company or detachment. They can be located in the BSA near the BSMC/FSMC when it is at a stable base camp. The Soldier restoration center may receive *hold* cases that are transferred from the first-line medical companies, as well as Soldiers from nearby units. Second-line Soldier restoration has more equipment and a greater range of COSC expertise. This permits a 72-hour holding capacity for a stable well-organized Soldier restoration center and may provide full stabilization.

THIRD-LINE SOLDIER RESTORATION

10-15. In some scenarios, units with Soldiers in need of Soldier restoration may be significantly closer to a Level III MTF such as a CSH than to a medical company or CSC unit. On order, a Soldier restoration program may be conducted by the CSH specialty clinic NP staff, which may also be augmented with personnel from a medical company/medical detachment, CSC. In such cases, a Soldier restoration program may be conducted by the CSH specialty clinic NP staff. It may be augmented with personnel from the medical company or detachment, CSC. Soldier restoration near a CSH must be kept clearly separate from the patient ward. It may be done at a MCW or in separate tents.

RETURN TO DUTY OF RECOVERED COMBAT AND OPERATIONAL STRESS REACTION SOLDIERS

10-16. Most COSR symptoms do not necessarily improve completely while the prospect of combat continues. The Soldiers should be given the positive expectation that they will RTD. Every possible effort should be made to return Soldiers to their original unit. The RTD of Soldiers following Soldier restoration depends on how near the Soldier's unit is, the availably of a means of transportation, and the tactical situation. Ideally, their units are contacted to send someone to get the Soldier or he may be returned to his unit by way of the personnel replacement company. Mental health personnel coordinate directly with the unit to which the Soldier is returning and/or with COSC, other medical personnel, or UMT supporting the unit's AO. These contacts can consult with the leaders of the Soldier's unit and facilitate the Soldier's acceptance back into his unit.

DOCUMENTATION

10-17. Soldiers receiving Soldier restoration interventions must be tracked from the initial contact until they are returned to their unit. A record must be maintained of interventions and activities provided during Soldier restoration, as well as the Soldier's response. A notation of the dates, any pertinent medical data, and providing unit is entered in the Soldier's medical record. An administrative summary of Soldier restoration services may be developed, stored, and disposed of as directed by the Soldier restoration center SOP. If the Soldier received BH treatment, documentation should be maintained as delineated in AR 40-66. A statistical record is maintained and sent to higher medical headquarters according to SOP.

Chapter 11

Behavioral Health Treatment

SECTION I — BEHAVIORAL HEALTH CARE

TREATMENT FOR BEHAVIORAL DISORDERS

11-1. Behavioral health treatment exists when there is an explicit therapist-patient or therapist-client relationship. Behavioral health treatment is provided for Soldiers with behavioral disorders to sustain them on duty or to stabilize them for referral/transfer. This is usually brief, time-limited treatment as dictated by the operational situation. Behavioral health treatment includes counseling, psychotherapy, behavior therapy, occupational therapy, and medication therapy. Treatment assumes an ongoing process of evaluation, and may include assessment modalities such as psychometric testing, neuropsychological testing, laboratory and radiological examination, and COSC providers' discipline-specific evaluations.

COMBAT AND OPERATIONAL STRESS CONTROL BEHAVIORAL HEALTH PROVIDER

11-2. There are five professional disciplines and two enlisted specialties that serve as BH/COSC providers, see Chapter 3. Army BH/COSC providers are officers who hold active individual clinical privileges granted by an MTF according to AR 40-68. Providers also include enlisted medical specialist who do not hold individual clinical privileges, but who work under direct supervision of a professional COSC provider. Behavioral health treatment is provided within the scope of the providers' clinical privilege, training, and experience.

SECTION II — BEHAVIORAL HEALTH TREATMENT PROTOCOLS AND MEDICATIONS

BEHAVIORAL HEALTH TREATMENT FOR SOLDIERS

11-3. Behavioral health treatment is provided to Soldiers with diagnosed behavioral disorders (see Chapter 8), and who require more intentions for their diagnoses. It is both inappropriate and detrimental to treat Soldiers with COSR as if they are a BDP. A therapeutic relationship may promote dependency and foster the "patient" role. Likewise, medication therapy and the highly structured treatment modalities imply the "patient" role. Medication for transient symptom relief (insomnia or extreme anxiety) may not be detrimental if there is no expectation that medication will continue to be prescribed.

BEHAVIORAL HEALTH PROTOCOLS AND PROCEDURES FOR TREATMENT

11-4. Behavioral health treatment is provided within a context of preventing symptoms of behavioral disorder from progressing in severity and improving the level of functioning. The principles of BICEPS apply. Treatment should not impede activities for other prevention levels (for example, universal, selective, and indicated). Treatment should not interfere with the Soldier's duties or the unit's mission. The COSC provider should emphasize Soldier's identity as a Soldier despite having a diagnosis. Transportation of the Soldier to the provider (or vice versa) and ability to assure reliable meeting times can be limiting factors, depending on the operational environment. Medication refills must be coordinated for availability.

Standards of Treatment

11-5. Treatment standards are the same in the deployed environment as in garrison. When operational requirements dictate that clinical standards of treatment/care are waived or relaxed, it must be approved by the AO COSC consultant. (AR 40-68 provides additional information.) Treatment should be tailored to the anticipated availability of the Soldier and COSC provider. Short-term interventions are more practical than long-term commitments. If longer-term treatment is necessary, design the intervention in time-limited modules. Under no circumstances should treatment diminish the Soldier's ability to provide self-care and to defend himself. Exceptions include emergency stabilization and preparation for evacuation. In addition, the VHA/DOD Clinical Practice Guidelines website (http://www.oqp.med.va.gov/cpg/cpg.htm) offer clinicians evidence-based assessment and treatment algorithms for acute stress disorder, PTSD, and many other behavioral disorders.

Army Regulations Governing Evaluations

11-6. Fitness for duty evaluations are conducted as necessary within the priorities of the supported commanders according to AR 40-501; psychiatrists should not initiate a medical evaluation board without first ensuring the Soldier has received adequate treatment. This treatment may not be available in theater. Command-directed evaluations are conducted as necessary within the priorities of the supported commanders according to DODD 6490.1 and MEDCOM Regulation 40-38. Clinical documentation should be safeguarded according to AR 40-66 and local command policy. Treatment should be conducted in a location that is as private as possible. Information can be released to a third party if the Soldier consents. Combat and operational stress control providers need to notify command when the Soldier's safety is in question (suicidal, homicidal) or if the Soldier is removed from his unit for medical observation. Additional release of information to command is on a need-to-know basis. For any questions on release of information on COSR and NP patients, consultation with supporting JAG office is advised. Also, maintaining the Soldiers health record with clinical data is required and accomplished according to AR 40-66 and supplemental theater policy as appropriate. Treatment should occur throughout the evacuation process and follow-up is expected at home station.

Chapter 12
Reconditioning

SECTION I — RECONDITIONING PROGRAM

LOCATION

12-1. Reconditioning programs are intensive efforts to restore those Soldiers triaged as a *refer* case, but who still have good potential for RTD. Referral to reconditioning can be from Level II MTF, COSC Soldier restoration program, or from Level III MTF. Reconditioning includes the rehabilitation of Soldiers with mental disorders, such as substance abuse/dependency. Reconditioning programs are conducted by COSC personnel and Soldiers usually participate for up to 7 days. Soldier participation may be extended by a case-by-case exception to theater evacuation policy. Reconditioning is conducted only in a theater where there are adequate COSC elements and supplies. When COSC resources are needed for other functional areas, any ongoing reconditioning program reduces its scope of services or closes. Reconditioning is provided at first-, second-, or third-line centers determined by location and available resources.

PROVIDERS

12-2. Conducting reconditioning programs is a mission of the medical company, CSC and the medical detachment, CSC (MRI). Reconditioning can also be done on a small-scale by a CSC medical detachment (MF2K) under some circumstances.

SECTION II — TENETS AND PROCEDURES OF RECONDITIONING

RECONDITIONING

12-3. Reconditioning may be considered an extension of Soldier restoration. Reconditioning is similar to Soldier restoration, but with potentially longer stay, treatment strategies focus on preventing atrophy of skills and assisting Soldiers in regaining skills needed for duty. Reconditioning makes more use of BH treatment modalities.

STRUCTURED MILITARY ENVIRONMENT TO SUSTAIN SOLDIER'S IDENTITY AS A SOLDIER

12-4. Like Soldier restoration, reconditioning emphasizes a highly structured military unit environment and schedule of activities in order to keep the Soldier from adopting a "patient" role. Maintaining a military environment is even more critical when colocated with hospitals or other service support elements.

REPLENISHMENT OF PHYSIOLOGIC STATUS AND CONFIDENCE

12-5. Reconditioning initially emphasizes physical replenishment and hygiene, but later shifts the emphasis to more closely match the conditions that the Soldier should expect when RTD. In order to restore confidence in the Soldier, unit formations are held on a regularly scheduled basis, and include the Soldiers receiving reconditioning.

OCCUPATIONAL THERAPY

12-6. Therapeutic occupations may include—

- Activities of daily living (for example, hygiene, physical fitness, and uniform maintenance).

- Educational activities (for example, common and collective Soldier task training, and life skills training).
- Militarily relevant work details according to the current functional level and MOS/duties of the Soldier.
- Leisure/recreational activities (for example, enjoyable, relaxing activities, including cooperative physical and mental activities, and basic relaxation techniques).
- Social participatory activities (for example, games and ceremonies).
- Group training in relaxation techniques.
- Ventilation and coaching where COSC (or medics trained by COSC personnel) encourage discussion about stressors and their impact on mental well-being. This technique helps the Soldier to restore personal perspective with questions and coaching. Individual counseling and therapy may improve the Soldier's functions as well.

12-7. When reclassification recommendations are considered, an adjacent CS or CSS unit may be able to provide a job that will match the Soldier's abilities. This provides an opportunity to demonstrate the Soldier's abilities and build confidence. The reconditioning personnel may recommend to S1/G1 that the Soldier be reclassified to another MOS.

FIRST-LINE RECONDITIONING PROGRAM

12-8. First-line reconditioning programs in the corps are staffed by task-organized CSC elements from the CSC company or by an MRI CSC detachment. Reconditioning can be conducted by an MF2K CSC detachment, but would preempt its Soldier restoration capability. If the inpatient NP workload is light, a small reconditioning center may be staffed by personnel from the NP ward and consultation service of an MF2K CSH. The first-line reconditioning center usually collocates with a CSH but must maintain its separate, nonhospital identity. It should not be situated among the hospital wards, or near the morgue, triage area, or helicopter pad. Within a theater/AO, the preferred option is to have one or two reconditioning centers placed such as to allow easy access from Soldier restoration centers. Under major, prolonged combat conditions, it may be more appropriate to have one reconditioning center behind each division. The reconditioning center is dependent on the hospital for support. The reconditioning center works with the supporting hospital by sending work parties of Soldiers in the program to assist in food preparation and delivery and cleanup chores. These Soldiers may also be used for assisting with work details throughout the hospital, but must be under direct supervision of either hospital or BH personnel. The reconditioning center uses the supporting hospital's medical records section to maintain the permanent case records. Cases in the reconditioning center are counted as patients in the reconditioning center on the daily hospital census. The cases are not counted as occupied beds when reporting the hospital bed occupancy. Upon disposition from the reconditioning center, whether for RTD, retraining for other duty, or evacuation, the reconditioning center prepares the chart for further evacuation or writes the discharge summary and closes the hospital's chart.

12-9. Some reconditioning cases will be able to return to far forward CS or CSS duty. However, many of the Soldiers who need reconditioning will be unable to return to their original unit, due to combat operations or the nature of their symptoms. Soldiers who undergo a 4- to 7-day Soldier restoration program in the corps should not be crossed off the division's personnel rolls (as specified in AR 40-216). Continuing contact with the unit increases RTD rates. For Soldiers who recover but are no longer on the division's rolls, every feasible effort should be made to return them to their original units. The chief, division MH and the CSC unit teams that are attached to reinforce the division should maintain frequent contact with the reconditioning centers that support the division. Coordination efforts, through the division support command (DISCOM) sustainment brigade and the personnel replacement system by division BH personnel, should attempt to facilitate return of recovered Soldiers to their original units. If recovered Soldiers cannot be returned to their previous small unit, consider forming them into cohesive pairs or small groups which can be reassigned to a new unit together. The standard corps evacuation policy is seven days, but some Soldiers with good potential for RTD may need a few more days at the reconditioning center. The corps evacuation policy is at the discretion of the theater commander. The principle purpose of a short evacuation policy is to keep hospital beds available for mass casualties and to minimize the expense and labor required to treat serious surgical and medical cases.

SECTION III — EVACUATION POLICY

EXTENDING THE THEATER EVACUATION POLICY

12-10. If need assessment indicates that a longer period of time (more than seven days) will achieve a Soldiers' RTD, the theater COSC consultant can request permission to extend the theater evacuation policy (up to 14 days). He submits the request for authority via the commander, MEDBDE, through the corps surgeon to the corps commander. The following facts apply: The reconditioning program, because of its austerity, is not a significant logistical burden to the corps. Soldiers in reconditioning perform useful work details and perimeter defense. Lastly, the increase in RTD is important in operations such as stability and reconstruction operations, where reducing personnel attrition becomes an important factor.

EVACUATION

12-11. Reconditioning cases that do not recover sufficiently to return to some duty within the designated evacuation period are evacuated from corps to EAC. They are best transported in cargo trucks and buses, bus ambulances, or an ambulance train. In the latter two cases, they should be assigned helper tasks. Use air evacuation only if there is no other alternative.

SECOND-LINE RECONDITIONING

12-12. Second-line reconditioning is conducted in the intermediate staging base outside the CZ in the theater. This center could be at a CSH, a fixed MTF, or a CSC unit element. The second line reconditioning center continues to emphasize physical fitness, Soldier skills, work details, and individual/group counseling/psychotherapy. Cases will be retrained for CSS duties at corps and theater levels. As soon as the recovering Soldiers are ready, the retraining site can shift to on-the-job training at a nearby CSS unit.

THIRD-LINE RECONDITIONING

12-13. Third-line reconditioning is for Soldiers with COSR and/or mental disorders who did not improve sufficiently at lower lines of reconditioning but still have RTD potential. These Soldiers are transferred to a reconditioning program at a regional MEDCEN or home station MEDDAC. These reconditioning centers, like the others, must maintain a military atmosphere and provide opportunities to engage in occupational therapy. Some Soldiers may also require retraining for other duty/MOS reclassification.

Appendix A

Combat and Operational Stress Control and Religious Support

ROLE OF UNIT MINISTRY TEAM

A-1. This appendix addresses the general role of the UMT in the commander's program for COSC. The UMT, imbedded within units down to battalion level, provide immediate support to leaders in fulfilling their COSC responsibilities. The UMTs also assist in training leaders to recognize combat and operational stress identification and intervention responsibilities. In cooperation with unit medical personnel, UMTs serve as a primary referral agency to BH resources.

RELIGIOUS SUPPORT FOR COMBAT AND OPERATIONAL STRESS CONTROL

A-2. Soldiers' inner resources are generally rooted in their religious and spiritual values. In combat, Soldiers often show more interest in their religious beliefs. When religious and spiritual values are challenged by the chaos of combat, Soldiers may lose connection with the inner resources that have sustained them. The UMT is the primary resource available to Soldiers experiencing such dilemmas and is a valuable resource in assisting them as they seek to refocus their spiritual values.

Unit Ministry Team Support for Combat and Operational Stress Control

A-3. The UMT provides preventive, immediate, and restorative spiritual, emotional support and care to Soldiers experiencing COSR.

Preventive Religious Support for Combat and Operational Stress Reaction

A-4. The UMT assists in preventing COSR and misconduct stress disorders through spiritual fitness training. Ministry of presence with Soldiers, assigned Department of Army civilians, and contractors is critical. The UMT provides a stabilizing influence on personnel and assists them in strengthening and regaining personally held spiritual values. Preventative activities include—

- Worship opportunities.
- Private and group prayer opportunities.
- Religious literature and materials.
- Scripture readings.
- Sacraments and ordinances.
- Assistance to personnel and families prior to deployment, emphasizing family strengths.

Immediate Religious Support for Combat and Operational Stress Reaction

A-5. The UMT assists commanders in the identification of personnel experiencing negative reactions to combat and operation stress, COSR, and misconduct stress behaviors. The UMT works closely with the unit's leaders and medical personnel to care for COSR cases through religious support and comfort. Immediate religious support activities may include—

- Conversation focused upon fears, hopes, and other feelings.
- Prayer for fallen comrades and memorial ceremonies and services.
- Rites, sacraments, and ordinances, as appropriate.
- Sacred scripture.

Restorative Religious Support for Combat and Operational Stress Control

A-6. Following an operation, a unit may require reconstitution. Surviving Soldiers may need to rebuild emotional, psychological, and spiritual strength. Depending upon the spiritual, emotional, and physical condition of the unit's Soldiers, the organic UMT may need augmentation from higher echelons or other units. Restorative religious support activities may include—

- Worship, sacraments, rites, and ordinances.
- Memorial ceremonies and services.
- Religious literature and materials.
- Grief facilitation and counseling.
- Reinforcement of the Soldiers' faith and hope.
- Opportunities for Soldiers to talk about combat experiences and to integrate those experiences into their lives.

Appendix B

Medical Detachment, Combat and Operational Stress Control

CONCEPT OVERVIEW

BACKGROUND

B-1. The current medical units, CSC in the Army inventory were initially developed over ten years ago. The MF2K redesign resulted in the development of the MRI. Recent events in the Global War on Terrorism in both Afghanistan and Iraq have indicated a need to update these units from MRI medical detachment, CSC to a multifunctional COSC detachment. The 2003 US Army Surgeon General's Mental Health Advisory Team (MHAT) report and observations indicate a need to update the unit in terms of modularity, multifunctionality, and its working relationship with support units and command surgeons from brigade to theater levels.

PROPOSED MEDICAL DETACHMENT, COMBAT AND OPERATIONAL STRESS CONTROL

B-2. The proposed medical detachment, COSC (TOE 08660G000) consists of a detachment headquarters, a main support section, and a forward support section. The main support section consists of its headquarters and an 18-Soldier BH team made up of social workers, clinical psychologists, psychiatrists, occupational therapists, psychiatric nurses, MH specialists, and OT specialists. The forward support section consists of an 18-Soldier BH team only. Each BH team is capable of breaking into six 3-person subteams, for battalion/company prevention and fitness support activities. This provides for a total of 12 subteams for each detachment, giving supported commanders more teams and more flexibility in the utilization of those teams, yet maintaining all of the capabilities of COSC.

Operational Environment

B-3. Soldiers will continue to experience the physical and mental impacts of high stress in both combat and stability and reconstruction operational environments. Advances in technology which will increase effectiveness, and the impact of the individual Soldier, will also increase dispersion and add to isolation and stress levels of Soldiers. The enemy of the future will look for methods that will have the greatest psychological impact on our future Soldiers; future enemies will not be our technological equals and will increasingly rely on terror and nontraditional methods to unnerve, injure, and demoralize both Soldiers and civilians. There is an identified need to update COSC units in terms of flexibility, modularity, multifunctionality, and its relationship to supported units and the surgeons responsible to commanders for the BH mission. The current configuration of the COSC units resulted in an uneven use of available COSC resources and that there was a greater need for preventive and limited fitness services forward. Commanders have communicated there is a greater need for active outreach and consultation.

Assumptions

B-4. Power projection will likely remain the fundamental strategic and operational imperative of our forces for the foreseeable future. The new medical detachment, COSC will be well suited to the future force, designed to move with and support the division and corps of the future force, and division and corps equivalents. The medical detachment, COSC will be assigned to a MEDBDE, MMB, or MEDCOM and will be operationally attached to supported units.

B-5. Each BCT has a MH section with one BH officer and one MH specialist and that all division and brigade MH sections are filled. The medical detachments, COSC will be able to augment BCTs MH sections. The medical detachment, COSC will require additional logistical, finance, maintenance, personnel, legal, FHP, and administrative services support.

Implication for Combat and Operational Stress Control

B-6. Future forces must be emotionally and mentally fit before deploying and resilient in battle in order to endure the fast pace and intense lethality of the battlefield. Commanders will need to be able to apply COSC capabilities at every level quickly and effectively. Reorganization of the medical detachment, CSC to a combined, multifunctional medical detachment, COSC will provide greater flexibility and greatly increase outreach and preventive services while maintaining capability for the entire spectrum of COSC.

Operational Concept, Mission

B-7. The medical detachment, COSC may be attached to a theater or division medical C2 headquarters in order to provide COSC casualty prevention, treatment, and management on an area basis.

B-8. The AUTL (see FM 7-15) linkage includes—

- ART 6.5. Provide Force Health Protection in a Global Environment.
- ART 6.5.1. Provide Combat Casualty Care.
- ART 6.5.1.5. Provide Mental Health/Neuropsychiatric treatment.
- ART 6.5.4. Provide Casualty Prevention.
- ART 6.5.4.5. Provide Combat Operational Stress Control Prevention.

Method of Employment

B-9. The medical detachment, COSC deploys with a division-sized force into a theater or with a task force of up to 25,000 Soldiers in stability and reconstruction operations. The medical detachment, COSC is attached for OPCON to the division/corps/theater medical C2 headquarters. The medical detachment, COSC is assigned to the MMB of units that have reorganized under the future force design. It is normally tasked by the command surgeon to support units within the command. The detachment commander as authorized by his medical C2 headquarters will coordinate operations with the command surgeon or the division/corps/theater psychiatrist. In the absence of a psychiatrist, the detachment commander will advise and consult with the command surgeon and the commander as required. Subteams may be task-organized and further attached for operational control to maneuver BCTs, and will operate in and forward from BSA, working closely with medical assets and combat religious support team (CRST). The detachment commander will coordinate efforts with the division and the CRST to provide in-theater prevention programs as required (suicide prevention, reunion, COSC training, TEM, and PTE stress management).

Detachment Headquarters

B-10. Plans, supervises and monitors the activities of the medical detachment, COSC in support of the COSC mission. See Table B-1 for a listing of personnel assigned to the detachment headquarters. The commander will establish a command post (CP) collocated with the supported unit surgeon (either division or sustainment brigade) and coordinate all activities of the detachment through the surgeon section. The detachment commander provides C2 and works closely with supported unit surgeon to provide logistics and other required services for the detachment. The field medical assistant and/or the detachment sergeant also establishes and coordinates logistics and required services support through the supported command staff for all detachment personnel and operations.

Table B-1. Detachment headquarters personnel

Detachment Commander (LTC/05, AOC 05A00)
Medical Operations Officer (CPT/03, AOC 70B67, MS)
Chaplain (CPT/03, AOC 56A00, CH)
Detachment Noncommissioned Officer in Charge (NCOIC) (SFC/E7, MOS 68X40)
Supply Sergeant (SGT/E5, MOS 92Y20)
Human Resources Specialist (SPC/E4, MOS 42A10)
Wheeled Vehicle Mechanic (SPC/E4, MOS 63B10)
Patient Administration Specialist (SPC/E4, MOS 68G10)
Religious Support Specialist (SPC/E4, MOS 56M10)
Cook (PFC/E3, MOS 92G10)
Note. The position of detachment commander may be filled by a (LTC, AOCs 73A00, 73B00, 60W00, 65A00 or 66C00). For description of duties, see Section I of the TOE and Department of Army (DA) Pamphlet 611-21.

Main Support Section

B-11. Included in the main support section is the detachment headquarters personnel identified in Table B-1 above and an 18-personnel BH team. See Table B-2 for list of personnel assigned to the BH team. This BH team can break into 6 three-person subteams, each with one officer, one NCO, and one specialist to conduct prevention and limited fitness operations in the battalion/company level in a maneuver BSA. Teams are collocated with a BCT medical unit. Modularized subteams may be combined and task-organized to provide or support any other COSC function, which includes larger scale Soldier restoration reconstitution, or any other COSC or BH mission as determined by its headquarters. Tasks for the main support section BH team includes:

- Consultation and coordination with commanders at team level (battalion and company).
- Coordination with CRST.
- Prevention and fitness support activities.
- Reconstitute any other COSC function as required.

B-12. When attached for support to any medical company Level II MTF, the grouped subteams are dependent on the supporting medical company for logistical support and quarters. The subteam can provide 3-day restoration for up to 50 Soldiers. With additional logistical support and quarters from supported command, the team can accommodate additional Soldiers under surge conditions.

Table B-2. Main section behavioral health team

Team Leader/Psychiatrist (MAJ/04, AOC 60W00)
Occupational Therapist (MAJ/04, AOC 65A00)
Social Work Officer (MAJ/04, AOC 73A67)
Behavioral Science Officer (CPT/03, AOC 67D00)
Clinical Psychologist (CPT/03, 73B67)
Psychiatric Nurse (CPT/03, 66C7T)
Occupational Therapy NCO (Four)—(Two SSGs/E6 and Two SGTs/E5, MOS 68W30/68W20)
Mental Health NCO (Four)—(One SSG/E6 and Three SGTs/E5, MOS 68X30/68X20)
Mental Health Specialist (Six)—(SPC/E4, MOS 68X10)
Note. For description of duties, see Section I of the TOE and DA Pamphlet 611-21.

Forward Support Section

B-13. This section consists of an 18-Soldier BH team. It is comprised of a section leader and five other BH officers, six MH/OT NCOs, and six MH specialists. This BH team can break into six three-person subteams, each with one officer, one NCO, and one specialist to conduct prevention and limited fitness operations in the BCT medical company in the BSA. Modularized subteams may be combined and task-organized to provide or support any other COSC function, which includes larger scale Soldier reconstitution, or any other COSC or BH mission as determined by its headquarters. The forward support section is dependent on the supported unit for C2 and logistical support. See Table B-3 for list of personnel assigned to the forward support section BH team.

Table B-3. Forward support section behavioral health team

Team Leader/Psychiatrist (MAJ/04, AOC 60W00)
Psychiatric Nurse (MAJ/04, 66C7T)
Clinical Psychologist (MAJ/04, 73B67)
Occupational Therapist (CPT/03, AOC 65A00)
Social Work Officer (CPT/03, AOC 73A67)
Behavioral Science Officer (CPT/03, AOC 67D00)
Occupational Therapy NCO (Two)—(One SSG/E6 and One SGT/E5, MOS 68W30/68W20)
Mental Health NCO (Four)—(One SSG/E6 and Three SGTs/E5S, MOS 68X30/68X20)
Mental Health Specialist (Six)—(SPC/E4, MOS 68X10)
Note. For description of duties, see Section I of the TOE and DA Pamphlet 611-21.

Organizational Concept

B-14. Required capabilities require the unit to—

- Deploy to power projection platform or point of debarkation and deploy with supported unit or deploy to theater and join supported unit.
- Provide all functions of COSC from company to theater level in close coordination with and under the direction of the supported unit surgeon.
- Provide prevention, assessment, treatment, and referral BH activities.
- Provide consultation with commanders from company to theater level.
- Conduct and coordinate administrative and logistical support to sustain operations.

Basis of Allocation Total Army Analysis

B-15. The basis of allocation is 0.333 per BCT; 1 per division; 2 per theater. In support of a theater, a COSC detachment provides support on an area basis and provides additional support to the division/corps on order. See Figure B-1 for breakdown of separate standard requirement codes (SRCs) of the medical detachment, COSC sections.

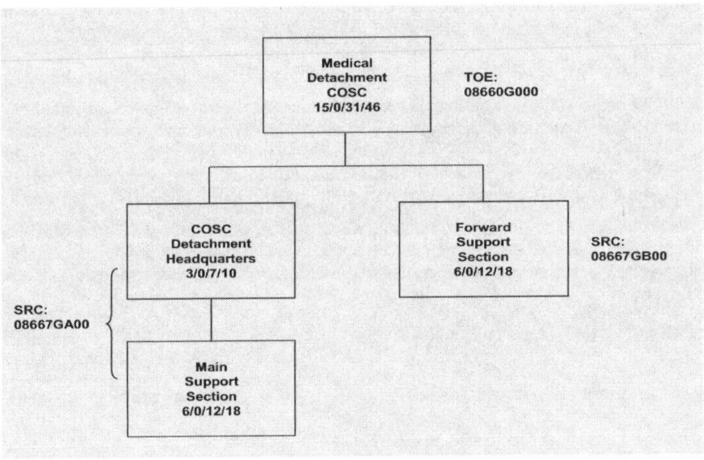

Figure B-1. Medical detachment, combat and operational stress control

Concept of Transition

B-16. Transition begins with the four active component detachments, one at a time. Personnel, transportation, and equipment changes will be minimal; transition should take a short amount of time and money to accomplish. Changes in the detachments are congruent with Force Modernization designs and personnel proponency issues. The proposed COSC detachment is designed with the Future Force in mind.

Patient Medical Records

B-17. The Surgeon General's MHAT report notes that there is a need to standardize BH reporting and documentation. The BH charts were inconsistently maintained and documentation did not always accompany patients through the evacuation chain. An additional finding was that there was no standardized method of collecting BH workload or clinical data and that no database that tracked evacuees provided reliable clinical information. The patient administration specialist (MOS 68G10) will supervise or perform functions as outlined in DA Pamphlet 611-21 for the Soldier caseload; maintain Soldier/patient accountability and records; interface with patient administration division of supported/collocated medical units.

Role of Religious Support Specialist

B-18. Findings of the MHAT include the need for chaplains to be aware of their role in COSC and that an aggressive chaplain outreach program should be executed. This requires that supervisory chaplains be involved in CRST integration with BH and primary care providers, and educate primary players in role of CRST in COSC. These activities require that the COSC chaplain be mobile and have an active presence in the maneuver brigades. This position is required to fulfill the full-time force protection requirements for the chaplain while in theater. The religious support specialist assists the COSC chaplain by screening individuals seeking counseling, coordinates for convoy security when the CRST goes out on site visits; assists chaplain with fund request; develops and coordinates needed religious support projects within the unit; coordinates with detachment for all vehicle services, communication system repairs, detachment training requirements and section accountability. This Soldier is cross-trained to develop skills for COSC.

Multifunctionality

B-19. The MH sections have a priority of prevention activities, but also have the capability to conduct or support any combat operational stress function as required. The COSC teams will be outfitted to a standard to gain the capability of prevention and limited fitness activities.

Modularity, Command and Control

B-20. The medical detachment, COSC attached early to supported units will have a better-integrated relationship. Detachment commanders and NCOICs will collocate with and coordinate all activities through supported unit surgeons. Detachments are designed to support a division in combat or a joint task force of up to 25,000 personnel in noncombat operations. The BH teams are designed to support and operate within maneuver BCTs in support of maneuver battalions for area coverage, collocating in BSA with the BSMC. Teams may be combined for task-organization to meet the entire spectrum of COSC as required.

Limitations

B-21. The medical detachment, COSC does not have its own logistics or services capability and must depend on the supported unit for logistical, legal, FHP, food service, finance, and personnel and administrative services. The COSC teams are small (one or two vehicles) and are dependent upon supported units for convoy and security operations. The logistical support necessary (space, tents, cots, and so forth) to provide Soldier restoration/reconstitution functions must be furnished by the supported element (division, corps, or theater).

Assumptions

B-22. Wartime activities will continue to cause physical and mental impacts of extremely high stress in both direct combat and the support operational environment. Advances in technology will impact the individual Soldier by increasing dispersion, and adding to their isolation and stress levels. The enemy of the future will look for methods that will have the greatest psychological impact on our future Soldier; future enemies will not be our technological equals and will increasingly rely on terror and nontraditional methods to unnerve, injure, and demoralize both Soldiers and civilians. Commanders will need to be able to quickly and effectively apply COSC capabilities at every level of the unit.

Appendix C

Medical Company and Medical Detachment, Combat Stress Control (Medical Force 2000)

SECTION I — MEDICAL COMPANY, COMBAT STRESS CONTROL (TOE 08467L000) (MEDICAL FORCE 2000)

MISSION

C-1. The mission of the AOE medical company, CSC is to provide comprehensive COSC support through directed interventions activities and COSC training for supported corps units. It provides DS to maneuver brigades lacking organic BH officers; augments units with BH assets; and provides area support. The medical company, CSC reconstitutes other COSC assets. The medical company, CSC provides COSC interventions and activities to indigenous populations as directed in stability and reconstruction operations, to include domestic support operations, humanitarian assistance, disaster relief, peace support operations, and detention facility operations. The medical company, CSC provides COSC interventions and activities to units in support of their readiness preparation and throughout their deployment cycle.

ASSIGNMENT

C-2. The AOE medical company, CSC (TOE 08467L000) is assigned to a corps MEDBDE (TOE 08422A100) or corps MEDCOM (TOE 08411A000).

EMPLOYMENT

C-3. Company headquarters usually locates with the corps medical headquarters. The employment of its teams includes their dispersal throughout the corps AO. Teams may be attached to an ASMB; a division or brigade medical company; a CSH; or other corps medical unit headquarters.

CAPABILITIES

C-4. At TOE Level 1, the medical company, CSC includes—

- Dividing the preventive section into six mobile COSC preventive teams.
- Dividing the restoration section into four mobile restoration teams with each team being equipped to hold 40 Soldiers at the same time. With additional logistical support, each team can accommodate additional Soldiers under surge conditions.
- Assisting higher headquarters BH staff with planning and coordination of COSC support, identifying the stress threat and mental and physical stressors, and the implementation of COSC functional areas.

DEPENDENCY

C-5. This unit is dependent on—

- Appropriate elements of the corps for legal; finance; field feeding; personnel and administrative services support; laundry and clothing exchange; mortuary affairs support; and security of EPW, detainee, and US prisoner patients.
- The medical headquarters to which it is assigned/attached for FHP; medical administration; logistics (including MEDLOG); medical regulating of patients; evacuation; coordination for RTD; and unit-level equipment and CE maintenance.

MOBILITY AND SECURITY

C-6. This unit is 100-percent mobile. It requires 100 percent of its organic personnel and equipment be transported in a single lift, using its organic vehicles. Upon relocation, Soldiers being held will require additional transportation. This unit is responsible for perimeter defense of its immediate operational area. However, it is dependent on appropriate elements of the corps for additional security, to include security of convoy operations. Personnel of the company (except the chaplain) are provided weapons for their personal defense and for the defense of their patients and/or held Soldiers.

ORGANIZATION

C-7. The AOE medical company, CSC (Figure C-1) is organized into a headquarters section, a preventive section with six (modular) CSCP, and a restoration section with four (modular) combat stress control restoration (CSCR) teams. The medical company, CSC is only present in the MF2K and is an RC organization that is programmed to be in the Army inventory until the year 2010. Under the MRI a new medical detachment, CSC, TOE 08463A000, will replace this unit.

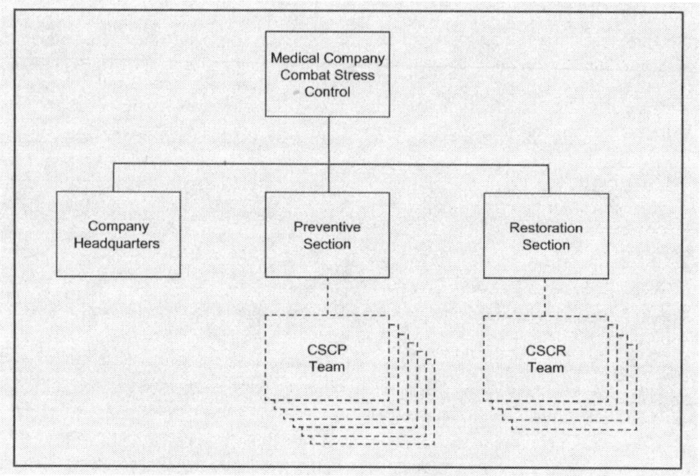

Figure C-1. Medical company, combat stress control

Company Headquarters Section

C-8. The company headquarters section provides C2 and unit-level administrative and maintenance support to its subordinate sections when they are collocated with the company. Personnel from the headquarters section are deployed with teams or task-organized COSC elements, as required. Personnel assigned to this section are identified in Table C-1.

Table C-1. Company headquarters section personnel 3/0/14

Company Commander (LTC/05, AOC 60W00, MC)
Chaplain (CPT/03, AOC 56A00, CH)
Field Medical Assistant (CPT/03, AOC 70B67, MS)
First Sergeant (1SG/E8, MOS 68W5M)
Mental Health NCO (SFC/E7, MOS 68X40)
Motor Sergeant (SSG/E6, MOS 63B30)
Supply Sergeant (SGT/E5, MOS 92Y20)
Equipment Record/Parts Sergeant (SGT/E5, MOS 92A20)
Wheeled Vehicle Mechanic (SGT/E5, MOS 63B20)
Human Resource Specialist (SPC/E4, MOS 42A10)
Nuclear, Biological, and Chemical Specialist (SPC/E4, MOS 74D10)
Wheeled Vehicle Mechanic (SPC/E4, MOS 63B10)
Unit Supply Specialist/Armorer (SPC/E4, MOS 92Y10)
Power-Generation Equipment Repairer (SPC/E4, MOS 52D10)
Cook (Three)— (Two SPC/E4 and One PFC/E3, MOS 92G10)
Note. The cook is deployed to assist the unit providing food service support to the company. He may participate in other CSC unit missions including outreach surveillance and stress control training.

Preventive Section

C-9. The 24-person preventive section staff is identified in Table C-2. The preventive section is task-organized to conduct COSC interventions and activities. Preventive section personnel may be task-organized with personnel of the COSC restoration section into teams for specific missions. The preventive section can divide into six CSCP teams. The section (and team) leader position may be held by any of the officers assigned to the section. The preventive section can augment or reconstitute the medical detachment, CSC teams.

Table C-2. Preventive section personnel 12/0/12

Psychiatrist (Three)—(MAJ/04, AOC 60W00)
Social Worker (Two)—(MAJ/04, AOC 73A67)
Psychiatrist (Three)—(CPT/03, AOC 60W00)
Social Worker (Four)—(CPT/03, AOC 73A67)
Team Chief (Six)—(SGT/E5, MOS 68X20)
Mental Health Specialist (Six)—(SPC/E4, MOS 68X10)

Restoration Section

C-10. The 44-person restoration section staff is identified in Table C-3. The restoration section may be task-organized to perform its COSC mission. Combat stress control restoration section personnel may also be task-organized with personnel of the CSCP section into teams for specific missions. This section can be divided into four CSCR teams. Each CSCR team can deploy a four-person mobile team using their HMMWV.

Table C-3. Restoration section personnel 12/0/32

Occupational Therapist (MAJ/04, AOC 65A00)
Psychiatric/Mental Health Nurse (Two)—(MAJ/04, AOC 66C00/7T)
Clinical Psychologist (MAJ/04, AOC 73B67)
Occupational Therapist (Three)—(CPT/03, AOC 65A00)
Psychiatric/Mental Health Nurse (Two)—(CPT/03, AOC 66C00/7T)
Clinical Psychologist (Three)—(CPT/03, AOC 73B67)
Mental Health NCO (Two)—(SFC/E7, MOS 68X40)
Occupational Therapy NCO (SSG/E6, MOS 68W30/N3)
Team Chief (Four)—(SSG/E6, MOS 68X30)
Patient Administration NCO (SGT/E5, MOS 68G20)
Occupational Therapy Sergeant (Two)—(SGT/E5, MOS 68W20/N3)
Mental Health NCO (Four)—(SGT/E5, MOS 68X20)
Patient Administration Specialist (Three)—(Two SPC/E4 and One PFC/E3, MOS 68G10)
Occupational Therapy Specialist (Five)—(SPC/E4, MOS 68W10/N3)
Mental Health Specialist (Ten)—(Six SPC/E4 and Four SPC/E3, MOS 68X10)

SECTION II — MEDICAL DETACHMENT, COMBAT STRESS CONTROL (TOE 08463L000) (MEDICAL FORCE 2000)

MISSION

C-11. The mission of the AOE MF2K medical detachment, CSC is to provide COSC interventions and activities to supported units in its AO. It augments division and brigade MH sections; provides direct support to combat brigades without organic BH officers; and provides area support in its AO. The medical detachment, CSC reconstitutes other brigade and division COSC assets. The medical detachment, CSC provides COSC interventions and activities to indigenous populations as directed in stability and reconstruction operations, to include domestic support operations, humanitarian assistance, disaster relief, peace support operations, and detention facility operations. The medical detachment, CSC provides COSC interventions and activities between deployments to units in support of their readiness preparation and throughout their deployment cycle. Under MRI this detachment will be replaced with medical detachment, CSC, TOE 08463A000.

> ***Note.*** The MF2K and MRI medical detachments, CSC conduct COSC support operations very similarly. However, in the MRI medical detachment, CSC, psychiatry assets are assigned to the restoration section and psychology assets are assigned to the preventive section whereas in the MF2K detachment, the manning strategy for psychology and psychiatry is reversed: psychologists are in the restoration section and psychiatrists are on the preventive teams.

ASSIGNMENT

C-12. The MF2K medical detachment, CSC (TOE 08463L000), is assigned to a corps MEDBDE (TOE 08422A100), or a corps MEDCOM (TOE 08411A000), or other medical task force C2 elements. Its teams may be attached to a medical company, CSC (TOE 08467L000); an ASMB (TOE 08456A000); a brigade or divisional medical company; or a CSH.

EMPLOYMENT

C-13. Detachment headquarters is usually located in the DSA. Its teams disperse and are employed throughout its AO.

CAPABILITIES

C-14. The MF2K medical detachment, CSC provides—

- Mobile COSC interventions from the preventive section which can divide into three preventive teams.
- Holding for 40 Soldiers by the restoration team for COSC interventions such as restoration. With additional logistical support, the team can accommodate additional Soldiers under surge conditions.

STAFF RESPONSIBILITIES

C-15. The medical detachment assists the C2 headquarters (to which it is assigned or attached) regarding planning and coordination of COSC support, stress threat, mental and physical stressors, stress behaviors, principles of COSC, and implementation of COSC functional areas.

DEPENDENCY

C-16. This unit is dependent on—

- Appropriate elements of the supporting unit for FHP; religious support; legal; finance; field feeding; personnel and administrative services support; laundry and clothing exchange; mortuary affairs support; and security of EPW, detainee, and US prisoner patients.
- Supporting unit for medical administration; logistics (including MEDLOG); medical regulating; evacuation; coordination for RTD; and unit-level equipment and CE maintenance.

MOBILITY AND SECURITY

C-17. This unit is 100-percent mobile. It requires 100 percent of its organic personnel and equipment be transported in a single lift, using its organic vehicles. Upon relocation, Soldiers being held will require additional transportation. This unit is responsible for perimeter defense of its immediate operational area. However, it is dependent on appropriate elements of the supporting unit for additional security, to include security of convoy operations. Personnel of the detachment are provided weapons for their personal defense and for the defense of their patients and/or held Soldiers.

ORGANIZATION

C-18. This 25-person unit (see Table C-4) is composed of a headquarters, a combat stress preventive section with three preventive teams, and a CSCR. The modular CSC teams found in the MF2K CSC medical detachment are similar to those found in the CSC medical company.

Table C-4. Detachment teams

Headquarters	*Preventive Section*	*Restoration Section*
	CSCP#1	CSCR
	CSCP#2	
	CSCP#3	

C-19. The detachment headquarters provides C2 for the detachment. It is responsible for planning, coordinating, and implementing COSC interventions and activities for supported units. It consists of three personnel: a detachment Commander (05, 60W00, MC), a detachment sergeant (E7, 68X40), and a wheeled-vehicle mechanic (E4, 63B10). The detachment commander also serves as a treating physician with the preventive section. The detachment NCOIC (a senior MH NCO) also serves as the restoration team sergeant. Detachment officers and NCOs from the prevention team and the restoration team may be assigned additional duties, which enhance the overall effectiveness of the headquarters section. See Table C-5 for personnel assigned to detachment headquarters.

Table C-5 Detachment headquarters personnel

Commander (LTC/05, AOC 60W00)
Detachment Sergeant (SFC/E7, MOS 68X40)
Wheeled Vehicle Mechanic (SPC/E4, MOS 63B10)
Note. The commander also serves as the psychiatrist in the preventive section.

Preventive Section

C-20. The 12-person preventive section staff is identified in Table C-6. The preventive section may be task-organized to conduct its COSC mission. Preventive section personnel may also be task-organized with personnel of the restoration section into teams for specific missions. The preventive section can divide into three preventive teams. The section (and team) leader position may be held by any of the officers assigned to the section. This section's COSC interventions and activities are—

- Unit needs assessment; consultation and education; critical event and transition management; COSC triage; stabilization (emergency); and BH treatment.
- Assisting with restoration and reconditioning at the CSC detachment program. Overseeing a 1- to 3-day COSC restoration program in a brigade, division, or ASMC holding section or in another area suitable for Soldiers experiencing COSR and/or other stress-related disorders.

Table C-6. Preventive section 7/0/6

Psychiatrist (LTC/05, AOC 60W00)
Clinical Psychologist (Three)—(MAJ/04, AOC 73B67)
Social Worker (Three)—(CPT/03, AOC 73A67)
Team Chief (Three)—(SGT/E5, MOS 68X20)
Mental Health Specialist (Three)—(SPC/E4, MOS 68X10)
Note. The psychiatrist is also counted in the headquarters section.

Restoration Team

C-21. The 9-person restoration team staff is identified in Table C-7. The restoration team is task-organized to provide COSC interventions and activities. Restoration team personnel may also be task-organized with personnel of the CSC preventive section into teams for specific missions. Each restoration team can deploy a four-person mobile team with a HMMWV. The CSCF usually collocates with a supported divisional medical company to provide mobile COSC support within a DSA and conduct restoration programs, as required. The CSCF provides staff and equipment for operating a restoration or reconditioning center. This section's COSC interventions and activities are—

- Unit needs assessment; consultation and education; critical event and transition management; COSC triage; stabilization; and BH treatment.
- Conducting Soldier restoration and reconditioning programs.
- Assisting the CSH psychiatric section when psychiatric ward capability is required (refer to FM 8-10-14).

Table C-7. Restoration team 3/0/6

Psychiatrist (MAJ/04, AOC 60W00)
Psychiatric/Mental Health Nurse (MAJ/04, AOC 66COO/7T)
Occupational Therapist (CPT/03, AOC 65A00)
Occupational Therapy NCO (SSG/E6, 68W30/N3)
Team Chief (SSG/E6, MOS 68X30)
Occupational Therapy Sergeant (SGT/E5, MOS 68W20/N3)
Mental Health NCO (SGT/E5, MOS 68X20)
Mental Health Specialist (Two)—(One SPC/E4 and One PFC/E3, MOS 68X10)

Glossary

SECTION I — ACRONYMS AND ABBREVIATIONS

AAR	after-action review
ABCA	American, British, Canadian, and Australian
ACR	armored cavalry regiment
AMEDD	Army Medical Department
AN	Army Nurse Corps
AO	area of operations
AOC	area of concentration
AOE	Army of Excellence
AR	Army regulation
ART	Army tactical task
ASMB	area support medical battalion
ASMC	area support medical company
AUTL	Army Universal Task List
BAS	battalion aid station
BCT	brigade combat team
BDP	behavioral disordered patient
BF	battle fatigue
BH	behavioral health
BICEPS	brevity, immediacy, contact, expectancy, proximity, and simplicity (See Section II for definition.)
BSA	brigade support area
BSB	brigade support battalion
BSMC	brigade support medical company
C	Centigrade
C2	command and control
C4I	command, control, communications, computers, and intelligence
CBRN	chemical, biological, radiological, and nuclear
CE	communications-electronics
CH	chaplain
CHL	combat health logistics
CHS	combat health support
COA	course(s) of action
CONUS	continental United States
COSC	combat and operational stress control (See Section II for definition.)
COSR	combat and operational stress reaction (See Section II for definition.)

CP	command post
CPT	captain
CRST	combat religious support team
CS	combat support
CSC	combat stress control
CSCF	combat stress control fitness
CSCP	combat stress control preventive
CSCR	combat stress control restoration
CSH	combat support hospital
CSR	combat stress reaction
CZ	combat zone
DA	Department of the Army
DD	Department of Defense
DEPMEDS	Deployable Medical Systems
DISCOM	division support command
DMHS	division mental health section
DNBI	disease and nonbattle injury(ies)
DOD	Department of Defense
DODD	Department of Defense Directive
DODI	Department of Defense Instruction
DS	direct support
DSA	division support area
EAB	echelons above brigade
EAC	echelons above corps
EAD	echelons above division
EMT	emergency medical treatment
EPW	enemy prisoner(s) of war
ETOD	end-of-tour debriefing
F	Fahrenheit
FAP	Family Advocacy Program
FHP	force health protection
1LT	first lieutenant
1SG	first sergeant
FM	field manual
FMC	United States Field Medical Card (Department of Defense Form 1380)
FRAGO	fragmentary order
FSB	forward support battalion
FSMC	forward support medical company
G-1	Assistant Chief of Staff (Personnel)
G-2	Assistant Chief of Staff (Intelligence)
G-3	Assistant Chief of Staff (Operations and Plans)

G-4	Assistant Chief of Staff (Logistics)
G-5	Assistant Chief of Staff (Plans and Policy)
G-9	Assistant Chief of Staff (Civil-Military Affairs)
GP	general purpose
HBCT	heavy brigade combat team
HHD	headquarters and headquarters detachment
HIP	help-in-place
HMMWV	high-mobility multipurpose wheeled vehicle
HSL	health service logistics
HSS	health service support
IED	improvised explosive device
JAG	Judge Advocate General
LTC	lieutenant colonel
MAJ	major
MC	Medical Corps
MCW	minimal care ward
MEDBDE	medical brigade
MEDCEN	medical center
MEDCOM	medical command (US Army)
MEDDAC	medical department activity
MEDLOG	medical logistics
METT-TC	mission, enemy, terrain and weather, troops and support available, time available, and civil considerations
MF2K	Medical Force 2000
mg	milligram(s)
MH	mental health
MHAT	Mental Health Advisory Team (US Army)
MMB	multifunctional medical battalion
MOS	military occupational specialty
MP	military police
MRI	Medical Reengineering Initiative
MRO	medical regulating officer
MS	Medical Service Corps
MSB	main support battalion
MSMC	main support medical company
MTF	medical treatment facility
MTOE	modification table of organization and equipment
NATO	North Atlantic Treaty Organization
NCO	noncommissioned officer
NCOIC	noncommissioned officer in charge
NP	neuropsychiatric

OEF	Operation Enduring Freedom
OIF	Operation Iraqi Freedom
OPCON	operational control
OPLAN	operation plan
OPORD	operation order
OPTEMPO	operational tempo
OSAT	operational stress assessment team
OSR	operational stress reaction
OT	occupational therapy
PA	physician assistant
PCP	phencyclidine hydrochloride
PD	psychological debriefing
PERSTEMPO	personnel tempo
PIES	proximity, immediacy, expectancy, and simplicity
POC	point of contact
PTE	potentially traumatizing event
PTSD	posttraumatic stress disorder
pub	publication
PVNTMED	preventive medicine
R&R	rest and relaxation
RC	Reserve Component
RTD	return to duty
S1	Adjutant (US Army)
S2	Intelligence Officer (US Army)
S3	Operations and Training Officer (US Army)
S4	Supply Officer (US Army)
SBCT	Stryker Brigade Combat Team
SGT	sergeant
SOI	signal operation instructions
SOP	standing operating procedure
SP	Army Medical Specialty Corps
SPC	specialist
SRC	standard requirement code
SSG	staff sergeant
TC	training circular
TDA	table of distribution and allowances
TEM	traumatic events management
TEMPER	tent, expandable, modular, personnel
TG	technical guide
TOE	table(s) of organization and equipment
TRADOC	United States Army Training and Doctrine Command

TSOP	tactical standing operating procedures
UCMJ	Uniform Code of Military Justice
UMT	unit ministry team
UNA	unit needs assessment
US	United States
USAMEDDC&S	United States Army Medical Department Center and School
USAMRMC	United States Army Medical Research and Materiel Command
USAR	United States Army Reserve
USMC	United States Marine Corps
VHA	Veterans Health Adminstration

SECTION II — TERMS

brevity, immediacy, contact, expectancy, proximity, and simplicity

An acronym used for the management of combat and operational stress reactions—*brevity* (usually less than 72 hours); *immediacy* (as soon as symptoms are evident); *contact* (chain of command remains directly involved in the Soldiers recovery and return to duty), *expectancy* (combat stress control unit personnel expectation that casualties will recover); *proximity* (of treatment at or as near the front as possible); *simplicity* (the use of simple measures such as rest, food, hygiene, and reassurance). Also known as BICEPS.

combat and operational stress control

A coordinated program for the prevention of and actions taken by military leadership to prevent, identify, and manage adverse combat and operational stress reactions in units. Also known as COSC.

combat and operational stress reaction

The expected, predictable, emotional, intellectual, physical, and/or behavioral reactions of Service members who have been exposed to stressful events in combat or military operations other than war. Also known as COSR.

5 R's

Actions used for combat and operational stress reaction control that include—**R**eassure of normality; **R**est (respite from combat or break from the work); **R**eplenish bodily needs (such as thermal comfort, water, food, hygiene, and sleep); **R**estore confidence with purposeful activities and contact with his unit; **R**eturn to duty and reunite Soldier with his unit.

reconditioning program

An intensive 4- to 7-day program (may be extended by exception to theater evacuation policy) of replenishment, physical activity, therapy, and military retraining for combat and operational stress control casualties and neuropsychiatric cases (including alcohol and drug abuse) who require successful completion for return to duty or is evacuated for further neuropsychiatric evaluation.

Soldier restoration

A 24- to 72-hour (1- to 3-day) program in which Soldiers with combat and operational stress reactions receive treatment.

stabilization

The initial short-term management and evaluation of severely behaviorally disturbed Soldiers caused by an underlying combat and operational stress reaction, behavioral health disorder, or alcohol and/or drug abuse reaction.

References

SOURCES USED

These are the sources quoted or paraphrased in this publication.

DEPARTMENT OF DEFENSE

These documents are available online at: http://www.dtic.mil/whs/directives/

DODD 6490.1, *Mental Health Evaluations of Members of the Armed Forces*, 1 October 1997

DODD 6490.2, *Comprehensive Health Surveillance*, 21 October 2004

DODD 6490.5, *Combat Stress Control (CSC) Program*, 23 February 1999

DODI 6490.4, *Requirements for Mental Health Evaluations of Members of the Armed Forces*, 28 August 1997

MULTISERVICE PUBLICATION

FM 6-22.5/MCRP 6-11C/NTTP 1-15M, *Combat Stress*, 23 June 2000

ARMY PUBLICATIONS

These publications are available online at: https://akocomm.us.army.mil/usapa

AR 40-216, *Neuropsychiatry and Mental Health*, 10 August 1984

FM 7-15, *The Army Universal Task List*, 31 August 2003

FM 22-51, *Leaders' Manual for Combat Stress Control*, 29 September 1994

DOCUMENTS NEEDED

These documents must be available to the intended users of this publication.

DEPARTMENT OF DEFENSE FORMS

These forms are available online at: https://akocomm.us.army.mil/usapa

DD Form 1380, *US Field Medical Card*

DD Form 2795, *Pre-Deployment Health Assessment*

DD Form 2796, *Post-Deployment Health Assessment*

MULTISERVICE PUBLICATION

AR 190-8/OPNAVINST 3461.6/AFJI 31-304/MCO 3461.1, *Enemy Prisoners of War, Retained Personnel, Civilian Internees, and Other Detainees*, 1 October 1997

ARMY PUBLICATIONS

These publications are available online at: https://akocomm.us.army.mil/usapa

AR 25-1, *Army Knowledge Management and Information Technology Management*, 15 July 2005

AR 40-3, *Medical, Dental, and Veterinary Care*, 12 November 2002

AR 40-61, *Medical Logistics Policies*, 28 January 2005

AR 40-66, *Medical Record Administration and Health Care Documentation*, 20 July 2004

AR 40-68, *Clinical Quality Management*, 26 February 2004

AR 40-400, *Patient Administration*, 12 March 2001

AR 40-501, *Standards of Medical Fitness*, 1 February 2005

AR 340-21, *The Army Privacy Program*, 5 July 1985

DA Pamphlet 611-21, *Military Occupational Classification and Structure*, 31 March 1999

FM 4-02, *Force Health Protection in a Global Environment*, 13 February 2003

FM 4-02.6, *The Medical Company—Tactics, Techniques, and Procedures*, 1 August 2002 (Change 1, 9 April 2004)

FM 4-02.7, *Health Service Support in a Nuclear, Biological, and Chemical Environment—Tactics, Techniques, and Procedures*, 1 October 2002

FM 4-02.10, *Theater Hospitalization*, 3 January 2005

FM 4-02.16, *Army Medical Information Management—Tactics, Techniques, and Procedures*, 22 August 2003

FM 4-02.17, *Preventive Medicine Services*, 28 August 2000

FM 4-25.12, *Unit Field Sanitation Team*, 25 January 2002

FM 8-10-6, *Medical Evacuation in a Theater of Operations—Tactics, Techniques, and Procedures*, 14 April 2000

READINGS RECOMMENDED

These sources contain relevant supplemental information.

JOINT OR MULTISERVICE PUBLICATIONS

Most joint publications are available online at: http://www.dtic.mil/doctrine/

Joint Pub (JP) 1-02, *Department of Defense Dictionary of Military and Associated Terms*, 12 April 2001 (As amended through 9 June 2004)

JP 3-0, *Doctrine for Joint Operations*, 9 September 2001

JP 4-02, *Doctrine for Health Service Support in Joint Operations*, 30 July 2001

FM 1-02/MCRP 5-12A, *Operational Terms and Graphic*, 21 September 2004

FM 4-25.11/NTRP 4-02.1.1/AFMAN 44-163(I)/MCRP 3-02G), *First Aid*, 23 December 2002 (Change 1, 15 July 2004)

FM 8-9/NAVMED P-5059/AFJMAN 44-151V1V2V3, *NATO Handbook on the Medical Aspects of NBC Defensive Operations AMedP-6(B), Part I—Nuclear, Part II—Biological, Part IIII—Chemical*, 1 February 1996

FM 8-284/NTRP 4-02.23(NAVMED P-5042)/AFMAN (I) 44-156/MCRP 4-11.1C, *Treatment of Biological Warfare Agent Casualties*, 17 July 2000 (Change 1, 8 July 2002)

FM 8-285(4-02.285)/NAVMED P-5041/AFJMAN 44-149/FMFM 11-11, *Treatment of Chemical Agent Casualties and Conventional Military Chemical Injuries*, 22 December 1995

FM 21-10/MCRP 4-11.1D, *Field Hygiene and Sanitation*, 21 June 2000

ARMY PUBLICATIONS

Most of these publications are available online at: https://akocomm.us.army.mil/usapa

AR 71-32, *Force Development and Documentation—Consolidated Policies*, 3 March 1997

AR 635-200, *Active Duty Enlisted Administrative Separations*, 6 June 2005

FM 3-0, *Operations*, 14 June 2001

FM 3-06.11, *Combined Arms Operations in Urban Terrain*, 28 February 2002

FM 3-07, *Stability Operations and Support Operations*, 20 February 2003

FM 3-19.40, *Military Police Internment/Resettlement Operations*, 1 August 2001

FM 3-21.21, *The Stryker Brigade Combat Team Infantry Battalion*, 8 April 2003 (Change 1, 31 July 2003)

FM 3-21.31, *The Stryker Brigade Combat Team*, 13 March 2003

FM 3-90.3, *The Mounted Brigade Combat Team*, 1 November 2001

FM 5-0, *Army Planning and Orders Production*, 20 January 2005

FM 4-0, *Combat Service Support*, 29 August 2003

FM 4-02.4, *Medical Platoon Leaders' Handbook—Tactics, Techniques, and Procedures*, 24 August 2001 (Change 1, 18 December 2003)

FM 4-02.19, *Dental Service Support in the Theater of Operations*, 1 March 2001

FM 4-02.21, *Division and Brigade Surgeons' Handbook (Digitized)—Tactics, Techniques, and Procedures*, 15 November 2000

FM 4-02.24, *Area Support Medical Battalion— Tactics, Techniques, and Procedures*, 28 August 2000

FM 4.25.10, *Field Hygiene and Sanitation*, 20 June 2002

FM 8-10-14, *Employment of the Combat Support Hospital Tactics, Techniques, and Procedures*, 29 December 1994 (will be revised as FM 4-02.14)

FM 8-42, *Combat Health Support in Stability Operations and Support Operations*, 27 October 1997 (will be revised as FM 4-02.42)

FM 8-50, *Prevention and Medical Management of Laser Injuries*, 8 August 1990 (will be revised as FM 4-02.50)

FM 8-55, *Planning for Health Service Support*, 9 September 1994 (will be revised as FM 4-02.55)

FM 10-27-4, *Organizational Supply and Services for Unit Leaders*, 14 April 2000 (will be revised as FM 4-20.05)

FM 27-1, *Legal Guide for Commanders*. 13 January 1992 (will be revised as FM 1-04.1)

FM 27-10, *The Law of Land Warfare*, 18 July 1956 (Reprinted with basic including Change 1, 15 July 1976) (will be revised as FM 1-04.10)

FM 63-20, *Forward Support Battalion*, 26 February 1990 (will be revised as FM 4-93.20)

FM 63-21, *Main Support Battalion*, 7 August 1990 (will be revised as FM 4-93.21)

FM 100-9, *Reconstitution*, 13 January 1992 (will be revised as FM 4-100.9)

FM 100-17, *Mobilization, Deployment, Redeployment, and Demobilization*, 28 October 1992 (will be revised as FM 3-35)

FM 100-17-5, *Redeployment*, 29 September 1999 (will be revised as FM 3-35)

MEDCOM Regulation 40-38, *Command-Directed Mental Health Evaluation*, 1 September 2001

TC 25-20, *A Leader's Guide to After-Action Reviews,* 30 September 1993

TRADOC Pamphlet 600-22, *Leaders Guide for Suicide Prevention Planning,* 16 February 2005 (Available at http://www.tradoc.army.mil/publications.htm)

United States Army Center for Health Promotion and Preventive Medicine

Most of these publications are available online at: http://chppm-www.apgea.army.mil/dhpw/

USACHPPM TG 240, *Combat Stress Behaviors,* June 2004

USACHPPM TG 241, *Combat Operational Stress Reaction* (*COSR*) (*"Battle Fatigue"*), June 2004

USACHPPM TG 242, *Combat Operational Stress Reactions* (*COSR*) *Prevention: Leader Actions,* June 2004

USACHPPM TG 243, *Combat Stress Card,* May 1999

Tables of Organization and Equipment

TOE 08411A000, *Medical Command, Corps* (*MRI*)

TOE 8422A100, *Medical Brigade, Corps* (*MRI*)

TOE 08456A000, *Area Support Medical Battalion*

TOE 08457A000, *Area Support Medical Company*

TOE 08463A000, *Medical Detachment, Combat Stress Control* (*MRI*)

TOE 08463L000, *Medical Detachment, Combat Stress Control* (*MF2K*)

TOE 08467L000, *Medical Company, Combat Stress Control* (*MF2K*)

TOE 08660G000, *Medical Detachment, Combat Operational Stress Control* (*Force Development Update*)

Department of Veterans Affairs

Veterans Health Administration, *Clinical Practice Guidelines (Mental Health),* available online at: http://www.oqp.med.va.gov/cpg

Index

References are to paragraph numbers except where otherwise specified.

Printed in Dunstable, United Kingdom

91222492R00183